Stem Cells and Prostate Cancer

Scott D. Cramer
Editor

# Stem Cells and Prostate Cancer

Springer

*Editor*
Scott D. Cramer
Department of Pharmacology
University of Colorado
Anschutz Medical Campus
Aurora, CO, USA

ISBN 978-1-4939-0094-7 ISBN 978-1-4614-6498-3 (eBook)
DOI 10.1007/978-1-4614-6498-3
Springer New York Heidelberg Dordrecht London



Printed on acid-free paper

Springer is part of Springer Science+Business Media (www.springer.com)

# Preface

Prostate cancer is common and kills people every day. Recently, understanding of the biology of adult tissue stem cells and their cognate cancers has identified striking similarities in normal stem cells and tumor-initiating cells, or the so-called cancer stem cells. The defining properties of a stem cell are self-renewal and multilineage differentiation. Many cancers possess tumor-initiating cells with these properties. Several groups have been investigating these principles in the prostate from multiple perspectives. *Stem Cells and Prostate Cancer* is meant to synthesize current directions in research on prostate stem cells and prostate cancer tumor-initiating cells.

Similarities between normal prostate stem cells and prostate tumor-initiating cells, for instance the ability for self-renewal and multilineage differentiation, have focused attention on normal stem cell biology. There are now very good data, summarized in this book, which demonstrate self-renewal and multilineage differentiation of populations of adult prostate cells from both mouse and humans. Although few studies have taken these experiments to clonal resolution, the mounting evidence is for one or more stem cell populations in the adult prostate. There is controversy regarding multiple aspects of prostate stem cell biology: Is it the cell of origin of prostate cancer or is it a more differentiated prostate cell that "gains" more stem-like properties, is there one prostate stem cell or multiple stem cells in different compartments, and what is the role of stem cells in castrate-resistant disease? These concepts and more are addressed by leaders in the field of *Stem Cells and Prostate Cancer*. The potential significance of the prostate stem cell in prostate cancer development and in the etiology of castrate-resistant disease makes this area of high clinical and translational significance for basic, translational, and clinical scientists interested in disease models.

The topics covered in *Stem Cells and Prostate Cancer* range from hormonal control of the prostate stem cell, methods of identification and characterization of prostate stem cells and prostate tumor-initiating cells, the role of the stem cell niche in differentiation, the tumor microenvironment, targeting the stem cell for prevention, and the use of stem cell models for validating prostate cancer genetics. The authors and topics were chosen to represent the spectrum of research in prostate stem cells from some of the best in the field. Each chapter represents a unique view on prostate stem cells. In general, I have had a very light hand in editing these chapters so that the intent, tone, and perspective remain those of the contributing authors.

One underlying technique that is described in virtually all chapters is tissue recombination developed and refined by Jerry Cunha over several decades of pioneering research. In tissue recombination, fetal urogenital sinus mesenchyme, dissected from rodent embryos, is recombined with prostate epithelium and regrafted into a mouse host. This technique is described in multiple places in *Stem Cells and Prostate Cancer*. Originally these studies were used to demonstrate the instructive power of the mesenchyme in prostate epithelial development. Through multiple iterations of the model, this technique has guided our understanding of hormonal control of prostate development, endocrine targets in cancer, the contribution of tumor-associated fibroblasts to prostate cancer development and, most recently as described in this book, the use in evaluating prostate stem cells. The reader will find it clear that no definitive description of work on prostate stem cells is without discussion of the valuable contributions of tissue recombination to the field. My hat goes off to Jerry for his pioneering work that has facilitated our progress in the prostate stem cell field in uncountable ways.

Aurora, CO, USA Scott D. Cramer

# Contents

# Chapter 1
# Prostate Stem Cells, Hormones, and Development

**Gail S. Prins and Wen-Yang Hu**

**Abstract** While it is established that prostate cancer is a hormone-dependent disease, the cell(s) of origin of prostate cancer, i.e., the tumor-initiating cells, is still in debate. Strong evidence has emerged which indicates that prostate cancer can originate from both basal and luminal epithelial cell populations. In addition, prostate epithelial stem cells are candidates for the tumor-initiating cell based on work in hematopoietic and breast cancers and because of the growing acceptance of the cancer stem cell paradigm. To appreciate the interrelationships between the multiple cells of origin of prostate cancer, it may be necessary to first fully understand the prostate stem cell differentiation lineage during normal development and adult tissue maintenance as well as the factors that regulate stem cell self-renewal and lineage commitment. Recent advances in stem cell research have permitted isolation of prostate stem cells and shed light on the hierarchical relationship between the epithelial stem cells and their differentiated lineage. Furthermore, prostate cancer stem cells have been isolated and characterized from several prostate tumors which may provide an explanation for the known clinical and molecular heterogeneity of human prostate cancers. Although prostate stem cells and prostate cancer stem cells appear to be androgen receptor negative, new findings have established key roles for several other hormones in regulating prostate stem cells and their niche. Together, this new knowledge should allow for greater insight into the details of prostate development and to increased understanding of prostate cancer initiation and progression. In this chapter we will highlight recent advances in hormone modulation of prostate stem cells and their early progeny in development, normal tissue homeostasis, and cancer.

---

G.S. Prins, Ph.D. (✉) • W.-Y. Hu, Ph.D.
Department of Urology, University of Illinois at Chicago, 820 South Wood Street, (M/C955) Suite 132, Chicago, IL 60612, USA
e-mail: gprins@uic.edu

S.D. Cramer (ed.), *Stem Cells and Prostate Cancer*, DOI 10.1007/978-1-4614-6498-3_1, 

## 1.1 Prostate Gland Development

The prostate gland develops embryologically from the endodermal urogenital sinus (UGS) under the influence of androgens produced by fetal Leydig cells upon chorionic gonadotropin stimulation. In humans, prostate development occurs during the second and third trimester and is complete at the time of birth (Lowsley 1912; Prins 1993). The prostatic portion of the urethra develops from the *pelvic* (middle) part of the UGS, and prostate development initiates when UGS epithelium in this region penetrates into the surrounding mesenchyme to form the primordial prostate buds. The glandular epithelium of the prostate differentiates from these endodermal UGS cells and the associated mesenchyme differentiates into the prostate stroma, which primarily contains fibroblasts and smooth muscle cells (Prins 1993; Donjacour and Cunha 1993, 1995; Prins and Putz 2008; Moore and Persaud 2008). While the human prostate does not consist of separate lobes, four morphologically distinct prostatic zones have been identified by McNeal. The prostatic urethra divides the prostate into an anterior fibromuscular zone and a posterior glandular portion that contains the peripheral, central, and transition zones (McNeal 1983).

A significant amount of information about prostate gland development at the morphologic, cellular, and molecular levels has been derived from studies using rodent models. In contrast to humans, the rodent prostate gland is rudimentary at birth and undergoes the majority of its development during the first 15 days of life. Although the developmental process is continuous, the development of the rodent prostate can be categorized into five distinct stages involving determination, initiation or budding, branching morphogenesis, differentiation, and pubertal maturation (Prins and Putz 2008). Determination of the prostate occurs before there is clear morphological evidence of a developing structure and involves expression of molecular signals that commit a specific field of UGS epithelial cells to a prostatic cell fate. Development of the prostate phenotype commences as UGS epithelial stem and progenitor cells form outgrowths or buds that penetrate into the surrounding UGS mesenchyme (UGM) in the ventral, dorsal, and lateral directions caudal to the bladder (Cunha et al. 1983; Cunha 1973, 1976, 1984). At birth, the ventral, dorsal, and lateral rodent prostate lobes primarily consist of unbranched, solid, elongating buds or ducts, and subsequent outgrowth and patterning occur postnatally (Fig. 1.1a). During this time, proliferation of epithelial cells occurs primarily at the leading edge of the ducts (i.e., distal tips) (Prins et al. 1992). Branching morphogenesis begins when the elongating UGS epithelial buds contact the prostate mesenchymal pads that are peripheral to the periurethral smooth muscle. At that point, secondary, tertiary, and further branch points are established with continued outgrowth in the proximal-to-distal direction and with increased complexity (Hayashi et al. 1991; Timms et al. 1994). Epithelial and mesenchymal cell differentiation is temporally coordinated with branching morphogenesis. Lumenization of the solid epithelial cords begins in the proximal ducts and spreads to the distal tips, occurring concomitantly with epithelial differentiation into separate basal and luminal cell layers to form a simple columnar epithelium (Fig. 1.1b).

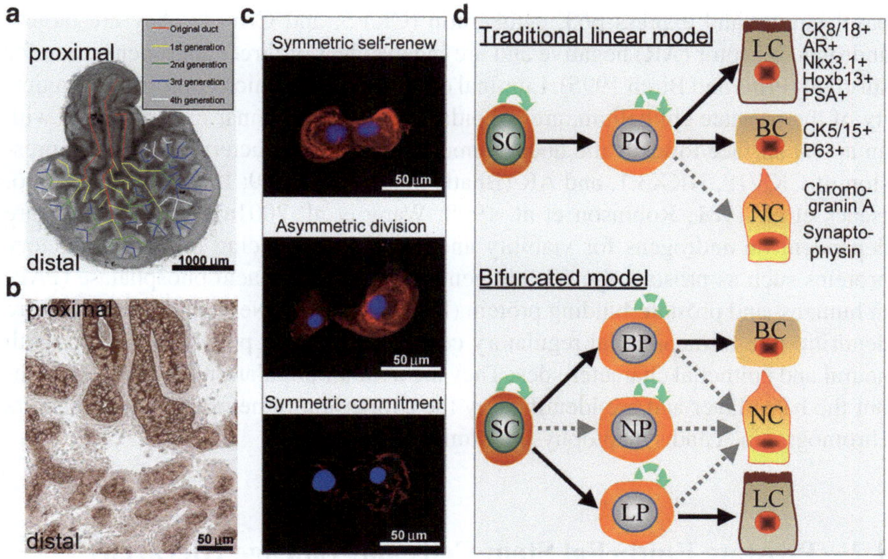


**Fig. 1.1** Prostate development and stem cells hierarchy. (**a**) Branching morphogenesis of rat ventral prostate lobe. The gland was removed at birth and cultured for 90 h as described (Huang et al. 2009). Images were taken every 30 min to track the branching events and a color-coded skeleton is used to indicate the generation of branches according to the following convention: The three primary buds that emerged from the UGS are considered the original ducts (*red*), branches that formed off the primary ducts are considered the first generation (*yellow*), and branches that formed off the first generation and elongate are considered the second generation (*green*), third generation (*blue*), and fourth generation (*white*). (**b**) Day 4 rat ventral prostate immunostained for CK8/18 reveals the proximal-to-distal spread of epithelial differentiation and lumen formation. Ducts in the proximal region of the gland have initiated lumenization concomitant with differentiation of luminal cells positive for CK8/18. In contrast, distal regions contain solid epithelial ducts with minimal luminal cell differentiation. (**c**) Immunolabeling of human prostate stem cells (isolated from day 4 prostaspheres) with stem cell marker CD49f identifies three types of stem cell divisions: (1) symmetric renewal, one stem cell gives rise to two identical daughter stem cells ($CD49f^{high}$); (2) asymmetric division, one stem cell gives rise to one daughter stem cell ($CD49f^{high}$) and one daughter progenitor cell ($CD49f^{low}$); and (3) symmetric commitment, one stem cell gives rise to two differentiated daughter cells ($CD49f^{low}$). (**d**) Prostate stem cell hierarchical models: in the traditional linear hierarchy model (*top*), self-renewing prostate stem cells (SC) give rise to intermediate, transit-amplifying progenitor cells (PC). These cells have high proliferative capacity and enter differentiation pathways to give rise to terminally differentiated luminal cells (LC), basal cells (BC), and neuroendocrine cells (NC). In the bifurcated model (*bottom*), a common SC with self-renewal capacity undergoes asynchronous cell division to give rise to lineage-restricted basal progenitor cells (BP) and luminal progenitor cells (LP). These PCs possess transient self-renewal capacity and terminally differentiate into basal and luminal epithelial cells. The lineage of neuroendocrine cells (NC) is unclear and may arise from the hierarchical intermediate BP and LP that produce BC and LC, or it may have a separate neuroendocrine progenitor (NP) origin

Mature prostate ducts contain three phenotypically and functionally distinct epithelial cell types (basal, luminal, and neuroendocrine cells) embedded in a fibromuscular stroma (Cunha et al. 1983, 1987; Cunha 1973; Long et al. 2005; Prins 1993; Prins and Putz 2008). Prostate basal epithelial cells are located adjacent to the basal lamina and express p63, cytokeratin (CK) 5, and CK 14. They are largely androgen receptor (AR) negative and are independent of direct androgen action for survival (Prins and Birch 1995). Luminal epithelial cells, which comprise the majority of the prostate epithelium, are cuboidal and short columnar exocrine cells with an apical surface towards the ductal lumen. They are characterized by the expression of CK8/18, NKX3.1, and AR (Bhatia-Gaur et al. 1999; Hayward et al. 1996; Isaacs et al. 1981; Robinson et al. 1998; Wang et al. 2001a). Luminal cells are dependent on androgens for viability and function, producing prostatic secretory proteins such as prostate-specific antigen (PSA), prostatic acid phosphatase (PAP) in humans, and prostate-binding protein (PBP) in rodents. Neuroendocrine cells are dendritic-like intraepithelial regulatory cells with a hybrid phenotype having both neural and epithelial characteristics. They are a minor population scattered throughout the basal layer and are identified by the expression of neuroendocrine markers chromogranin A and synaptophysin (Rumpold et al. 2002).

## 1.2 Prostate Epithelial Stem Cells and Lineage Hierarchy

Adult prostate stem cells have been identified in human and rodent prostate glands where they play an essential role in tissue replenishment throughout life (Bhatt et al. 2003; Garraway et al. 2010; Goldstein et al. 2008; Isaacs 2008; Kasper 2007; Lawson et al. 2010; Leong et al. 2008; Liu et al. 2011; Xin et al. 2007). This rare cell type self-renews and has potential to differentiate into the three distinct epithelial cell types, essential characteristics of bonafide stem cells. Prostate stem cells are relatively growth quiescent, occasionally dividing to self-renew and generate daughter progenitor cells. Studies across multiple systems as well as the prostate epithelium have characterized three types of stem cell divisions (Fig. 1.1c): (1) symmetric division, aka symmetric self-renewal, which generates two identical stem cells; (2) asymmetric division which generates a single self-renewing stem cell and a daughter cell that has entered the earliest stage of differentiation (progenitor cell); and (3) symmetric commitment division whereby a stem cell produces two daughter progenitor cells (Morrison and Kimble 2006; Scaffidi and Misteli 2011; Tomasetti and Levy 2010; Wu et al. 2007). Unlike the stem cell, the daughter progenitor cell has transit-amplifying capacity through rapid cell divisions. As the progenitor cell proliferative potential is exhausted, it undergoes terminal differentiation.

While prostate stem cells are present in several regions of rodent prostatic ducts, accumulated prostate stem cells with considerable growth potential have been found in the proximal region of ducts close to the UGS, and the survival of these cells does not require the presence of androgens (Tsujimura et al. 2002; Goto et al. 2006). Primitive proximal prostate stem cells that are able to regenerate functional prostatic tissue *in vivo* are also programmed to reestablish a proximal-distal ductal axis. In contrast, prostate stem cells in the distal region of the prostate duct have more limited growth potential and require androgens for survival (Goto et al. 2006). See Chapter 6 for a more complete discussion of the prostate stem cell niche.

Although the lineage hierarchy for the prostate epithelium has not been settled, epithelial differentiation of stem cells into differentiated basal, luminal, and neuroendocrine cells has been documented in the rodent prostate and in isolated human prostate cancer stem cells (Hudson 2004; Isaacs 2008; Kasper 2007; Long et al. 2005; Robinson et al. 1998; Wang et al. 2001a, b). These studies on stem cell differentiation have been observed with changing patterns of cytokeratins, cell-specific markers, and alterations in AR expression, an early marker of luminal epithelial cell differentiation. Two models have been proposed for the prostate epithelial stem cell lineage into differentiated basal, luminal, and neuroendocrine cell types (Fig. 1.1d). In the traditional linear hierarchy model, self-renewing prostate stem cells residing in the basal cell layer undergo asymmetric cell division giving rise to daughter progenitor cells with high proliferative potential, aka the transit-amplifying cells. In response to signals from the stem cell niche, these cells enter early differentiation pathways to eventually form separate basal, luminal, and neuroendocrine cells (Hudson 2004; Hudson et al. 2000; Isaacs and Coffey 1989). Phenotypic intermediate-type cells that co-express basal and luminal markers have been observed both *in vitro* and *in vivo* (Garraway et al. 2003; Long et al. 2005; Prins et al. 1995; Robinson et al. 1998). This suggests that basal and luminal cells are hierarchically related through common progenitor cells that give rise to differentiated basal cells and luminal cells. In the bifurcated model, basal cells and luminal cells represent separate epithelial cell lineages that originate from a common stem cell. These lineages may be sustained by intermediate transit-amplifying cells and/or lineage-restricted basal and luminal cell progenitors (Hudson 2004; Long et al. 2005; Wang et al. 2001a). The lineage of neuroendocrine cells is unclear. Neuroendocrine cells may arise from the hierarchical prostate epithelial stem and progenitor cells that produce basal and luminal cells, or they may have a separate progenitor cell origin as shown in the bifurcated model.

## 1.3 Prostate Stem Cell Isolation and Characterization

It is widely accepted that adult stem cells are involved in normal tissue maintenance throughout life while cancer stem cells support cancer growth (Presnell et al. 2002; Smith et al. 2007). Although the cell(s) of origin for prostate cancer may include luminal, basal, neuroendocrine, progenitor, and stem cells (Goldstein et al. 2010a, b; Kasper 2008, 2009; Wang and Shen 2011), it is increasingly evident that the resultant prostate cancers contain cancer stem cells that continuously seed and maintain tumor growth (Gu et al. 2007; Maitland et al. 2011). While conventional therapies for prostate cancer eradicate the majority of cells within a tumor, most patients with advanced cancer eventually progress to androgen-independent, metastatic disease that remains essentially incurable by current treatment strategies. Recent evidence has shown that cancer stem cells are a subset of tumor cells that appear to be therapy-resistant and are responsible for maintaining cancer growth which may be the underlying cause of disease relapse (Cocciadiferro et al. 2009;

Maitland and Collins 2008; Miki and Rhim 2008; Oldridge et al. 2012; Wang et al. 2012). Thus understanding the regulation of both normal stem cells and cancer stem cells may provide new insight into the origin and treatment of prostate cancer. Towards this end, identification and characterization of these rare cell populations has been a major research effort during the past decade with marked progress being realized utilizing flow cytometry and prostasphere culture (Garraway et al. 2010; Liu and True 2002; Xin et al. 2005).

Flow cytometry sorts cell populations by their specific cell surface CD markers, and expression profiles of CD markers have been extensively described for both normal and neoplastic prostate cell types (Liu et al. 2004; Liu and True 2002). Importantly, multiple CD molecules are enriched in prostate stem cells including Sca-1 (Xin et al. 2005), α2β1 integrin (Collins et al. 2001), CD133 (Richardson et al. 2003; Vander Griend et al. 2008), CD44 (Liu et al. 2004), CD117 (Leong et al. 2008), CD49f, and Trop2 (Garraway et al. 2010; Goldstein et al. 2008, 2010a). Combinations of antibodies specific for these markers have been used to isolate stem-like cells by FACS from dissociated prostate tissues or epithelial cell cultures, and their stem cell capabilities have been tested using various *in vivo* systems (Goldstein et al. 2008; Guo et al. 2012; Leong et al. 2008). An example of this approach using 2-channel flow cytometry from primary prostate epithelial cell cultures is shown in Fig. 1.2a. It is important to note, however, that there is no current consensus on the antigenic profile required for isolating a pure stem cell population from prostate epithelium by flow cytometry. The disadvantages of FACS include relative low cell yield when using multiple stem cell markers and cell damage following dissociation, labeling, and sorting.

Side-population analysis utilizing flow cytometry in combination with functional properties of stem cells is a convenient tool to characterize stem-like cells within mixed epithelial cell populations. Stem cell side-populations were first identified in hematopoietic stem cells enriched from heterogeneous cell populations based upon their unique ability to actively extrude Hoechst 33342 (Brown et al. 2007; Goodell et al. 1996). ABCG2 is a member of the family of ATP-binding cassette (ABC) transporters, and it pumps several endogenous and exogenous compounds out of cells including Hoechst 33342. Widely expressed in a variety of stem cells, ABCG2 was found to be a molecular determinant of the side-population phenotype and is recognized as a universal marker of stem-like cells (Ding et al. 2010; Zhou et al. 2001). The side-population assay, based on exclusion of Hoechst dyes, has proven to be a valuable method for identifying and sorting stem and early-stage progenitor cells in a variety of tissues and species. Application of this approach to assess putative prostatic stem cell numbers from heterogeneous prostate epithelial cell populations is shown in Fig. 1.2b (Bhatt et al. 2003; Brown et al. 2007; Mathew et al. 2009).

Another approach for enrichment and characterization of prostate stem cell populations is the prostasphere assay which utilizes a three-dimensional (3D) culture system to form spheroid structures (Garraway et al. 2010; Hu et al. 2011, 2012; Lukacs et al. 2010; Xin et al. 2007). First used for the isolation and characterization

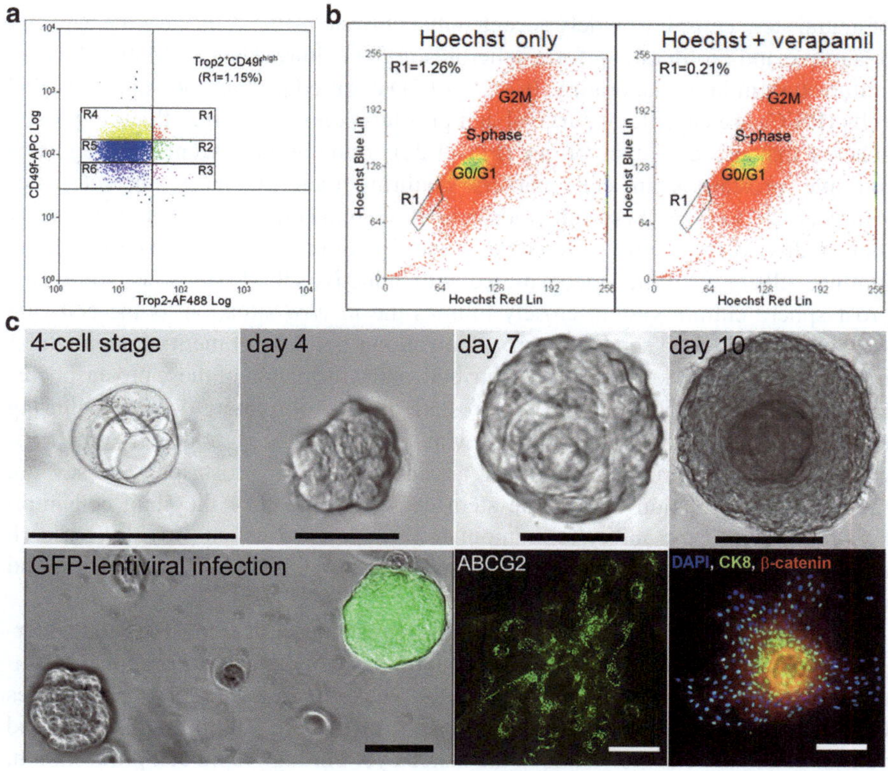


**Fig. 1.2** Methodological approaches for prostate stem cell isolation and characterization. (**a**) Flow cytometry analysis of normal human prostate epithelial cells (PrEC) from primary cell culture following labeling with Trop2-AF488 and CD49f-APC antibodies. A subpopulation of Trop2$^{+}$CD49f$^{high}$ (R1 = 1.15%, *red*) represents prostate stem cells. (**b**) Hoechst 33342 dye efflux fluorescence-activated flow cytometry analysis reveals a side population in human PrEC (gated as *R1*) that actively excludes the dye. PrEC were stained with 5 μg/mL of Hoechst 33343 either in the absence or presence of 50 μM of verapamil hydrochloride, an ABCG2 inhibitor. Windows for the side population are determined by comparison of cells without and with verapamil in each FACS analysis. (**c**) Prostate stem/progenitor cells isolation using prostasphere assay. Human PrEC from disease-free organ donors were established in primary 2D cultures and transferred to 3D Matrigel-slurry culture as described (Hu et al. 2011). Under these conditions, ~0.2–1% of primary PrEC cells (stem cell population) survive and undergo self-renew to form spheroid structures termed prostaspheres. By day 4 of culture, prostaspheres 30–40 μm in diameter are visible, increasing in size through transit amplification to 60–100 μm by day 7. To confirm clonality of the spheroids, mixed primary PrEC cells with or without lentiviral-GFP were transferred to 3D Matrigel cultures. At day 7, formed prostaspheres were either entirely GFP+ or GFP− (*bottom left*), indicating the clonal origin of prostaspheres. Day 7 prostasphere cells express multiple prostate stem cell markers (Hu et al. 2011) including the transporter protein ABCG2 (*bottom middle*). By day 10 of culture, prostaspheres grow >150 μm in diameter and form a visible double layer of cells (*top right*). Immunocytochemistry of a day 10 prostasphere shows central cells as differentiating CK8$^{+}$ (*green*) luminal cells (*lower right*) and peripheral cells p63$^{+}$ basal cells (Hu et al. 2011). Bar = 50 μm

of neural stem cells, it is widely accepted that only stem-like cells have the capacity to survive and proliferate to form spheroids in 3D culture. Using a Matrigel-slurry culture system in our laboratory, 0.2–1% of 2D cultured primary prostate epithelial cells (PrEC) are capable of survival and proliferation to form free-floating prostaspheres that are clonal in origin (Fig. 1.2c). Immunofluorescent labeling with multiple prostate stem cell markers confirms their stemness characteristics (Hu et al. 2011). That these spheroids consist of stem cells is best demonstrated by their ability to form fully differentiated and functional human prostate basal and luminal cells *in vivo* when reconstituted with inductive UGM (Hu et al. 2011). This prostasphere culture system closely mimics the *in vivo* situation as the cells are grown in a suspended semisolid gel, which allows the development of intercellular interactions. Several key variables contribute to the formation of these prostaspheres from PrEC including the age of the prostate donor, cell plating density, culturing techniques, and passage number, all of which influence the homogeneity or heterogeneity of the spheroids. The major advantages of the prostasphere assay are the functional isolation of prostate stem cells, expansion of the stem cell numbers *in vitro,* and the ability to manipulate them *in vitro* which provides research opportunities to identify regulation of stem and progenitor cell proliferation and differentiation.

At early stages of formation, the prostaspheres consist of stem-like cells undergoing synchronous self-renewal and asynchronous cell division to generate daughter early-stage progenitor cells that have not yet differentiated along cell lineages (Fig. 1.1c). By days 3–4 of culture, prostaspheres that are ~30 μm in diameter and consist of 20–40 cells are visible to the naked eye. Through rapid cell proliferation, they continue to grow with diameters reaching ~80–100 μm by day 7 (Fig. 1.2c). At this stage, cell markers indicate that the majority of cells express Nanog, Trop2, $CD49f^{high}$, ABCG2, CD133, CD44, and SSEA4 with no immunostaining for p63, CK8, NKX3.1, and HOXB13 suggesting that the day 7 spheroid cells consist of prostate stem and progenitor cell populations. Gene expression analysis of day 7 spheroids by real-time qRT-PCR supports their stem/progenitor cell status with lack of luminal cell gene expression (Fig. 1.3). Interestingly, although the cells are p63 negative by immunocytochemistry, p63 mRNA levels similar to parental PrEC cells are observed suggesting that early differentiation towards a basal cell lineage has initiated. With continued culture through day 10 to day 30, spheroid cells undergo cytodifferentiation, forming double-layered prostaspheres with sizes of 150–200 μm (Fig. 1.2c). Immunostaining of day 10 prostaspheres reveals that peripheral cells are p63-positive basal-type cells while centrally located cells are positive for CK8/18 and NKX3.1 indicating their differentiation towards a luminal phenotype (Fig. 1.2c). With continued culture under basal conditions through day 30, prostaspheres form branching-type structures and undergo functional differentiation as indicated by PSA gene induction (Hu et al. 2011). Furthermore, their growth and differentiation can be driven by various conditions including coculture with stromal cells and treatment with differentiating factors such as hepatocyte growth factor (HGF) (Schalken 2007) or hormones as described below.

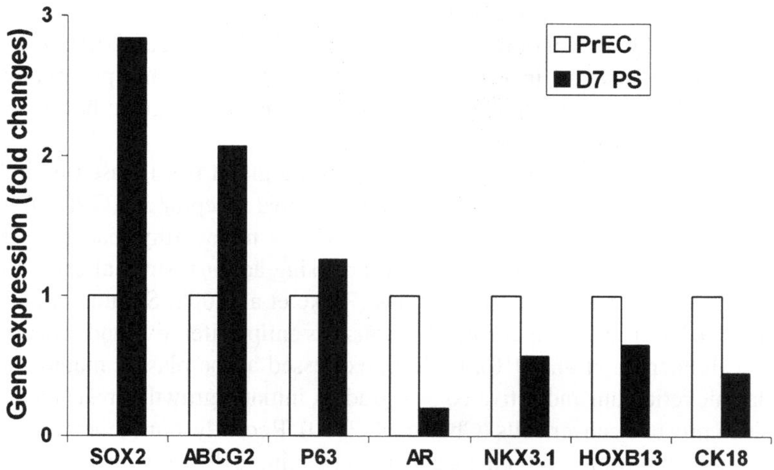


**Fig. 1.3** Day 7 Prostasphere gene expression analyzed by real-time qPCR. Relative to the normal parental PrEC cells in 2D primary culture, day 7 prostasphere cells express increased levels of prostate stem cell markers Sox2, ABCG2, and basal cell marker p63 and low to negligible (<30–35 Ct cycles) levels of luminal cell differentiation markers including AR, NKX3.1, HOXB13, and CK18

## 1.4 Hormone Receptor Expression and Hormonal Regulation of Prostate Stem and Progenitor Cell Self-Renewal and Differentiation

Androgens are essential for prostate gland development and maintenance throughout life and are believed to play central roles in prostate cancer initiation and progression. Despite this, prostate epithelial stem and early progenitor cells are AR negative (Hu et al. 2011; Kasper 2009; Oldridge et al. 2012) and are thus not directly regulated by androgen action. As a result, any effects of androgens on prostate epithelial stem cell homeostasis and differentiation are most likely mediated through indirect actions on the stem cell niche which includes $AR^+$ stromal cells and, in the mature prostate, $AR^+$ luminal epithelial cells (Berry et al. 2008). Androgens have been shown to influence the secretion of multiple paracrine-acting factors by these cells during prostate development and in the adult tissue that may influence the stem cell niche including *Fgf*s, *Shh*, and *Wnt*s (Prins and Putz 2008). While several studies have shown that prostate cancer stem-like cells are similarly AR negative (Kasper 2009; Oldridge et al. 2012), there are scattered reports on direct androgen action and AR protein in prostate cancer-initiating cells and prostate cancer stem cell subpopulations (Sharifi et al. 2008; Vander Griend et al. 2010). In addition to androgens, a number of other hormones are known to regulate prostate growth and function and to influence growth and progression of prostate cancer including estrogens (Prins and Korach 2008), retinoids (McCormick et al. 1999; Schenk et al. 2009), prolactin (Dagvadorj et al. 2007), growth hormone, and IGF-1 (Chan et al. 1998; Wang et al. 2005). Further, there is clear evidence that mammary gland stem cells and daughter progenitors are under direct regulation by several of these hormones (Asselin-Labat et al. 2010; Joshi et al. 2010). In this context, we investigated whether prostate stem and progenitor cells express other hormone receptors and respond to the non-androgenic hormones that are known to influence the prostate gland.

Past research on estrogen action in the prostate gland has focused entirely on estrogen receptor (ER)α, ERβ, and G protein-coupled receptor 30 (GPR30) within differentiated stromal, basal, and luminal cells. It is noteworthy that the different ERs within these cell types have apparent opposing actions; stromal cell ERα has proliferative and cancer-promoting actions (Ricke et al. 2008; Sissung et al. 2011) while ERβ in basal and luminal epithelial cells has antiproliferative and proapoptotic activity (McPherson et al. 2010). GPR30, expressed at the plasma membrane and endoplasmic reticulum and activated by estradiol, initiates growth arrest and induces necrosis in prostate cancer cells (Chan et al. 2010). Recently, our laboratory discovered that human prostate epithelial stem and progenitor cells from disease-free prostates express robust levels of ERα, ERβ, and GPR30 mRNA and protein (Hu et al. 2011). Further, prostaspheres cultured from primary PrEC in 1 nM estradiol-17β ($E_2$) exhibited a marked increase in spheroid size and number (Hu et al. 2011) with elevated expression of multiple stemness genes at day 7 of culture as compared to vehicle alone (Fig. 1.4a). Using a side-population analysis of primary PrEC, we noted a biphasic effect of estradiol with increased stem cell numbers at 1–10 nM $E_2$ but limited stimulation at higher doses (Fig. 1.4b). Taken together, these findings implicate prostate stem/progenitor cells as direct estrogen targets and indicate that estrogens support stem cell self-renewal and progenitor amplification and maintain their stemness state within the prostate gland. Moreover, these results raise the intriguing possibility that prostate stem and early progenitor cell populations may be susceptible targets of elevated estrogen levels in aging men (Vermeulen et al. 2002).

To evaluate whether prostate cancer stem-like cells may likewise express ERs and respond to estrogens, we examined ERα, ERβ, and GPR30 expression in stem and progenitor cells from prostate cancer specimens (Fig. 1.4c). Prostaspheres were cultured from primary prostate cancer cells (PCa-E) and matched benign prostate epithelial cells (EPZ) from the same patient at prostatectomy (kindly supplied by Dr. L Nonn, University of Illinois at Chicago). The PCa-E cells were cultured from pathologically confirmed cores containing >80% cancer cells and expressed significantly elevated AMACR and reduced NKX3.1 as compared to the EPZ cells. As shown in Fig. 1.4c, there was a sixfold increase in ERα mRNA and 8–12-fold increase in GPR30 expression in both the patient benign and cancerous prostasphere cells as compared to spheroids grown from normal donor PrEC. For ERβ, there was a fourfold increase in the PCa-E-derived stem/progenitor cells but not the benign EPZ cells relative to normal donor PrEC expression. Since the prostaspheres from PCa-E were mixed stem and progenitor cells that are not confirmed as prostate cancer stem-like cells, ERs were also evaluated in two human prostate cancer stem-like cell lines, HPET (Gu et al. 2007) and HuSLC (kindly supplied by Dr. S. Kasper, University of Cincinnati). Each cell line was generated from separate Gleason score 9 tumors, spontaneously immortalized and is capable of fully reestablishing the

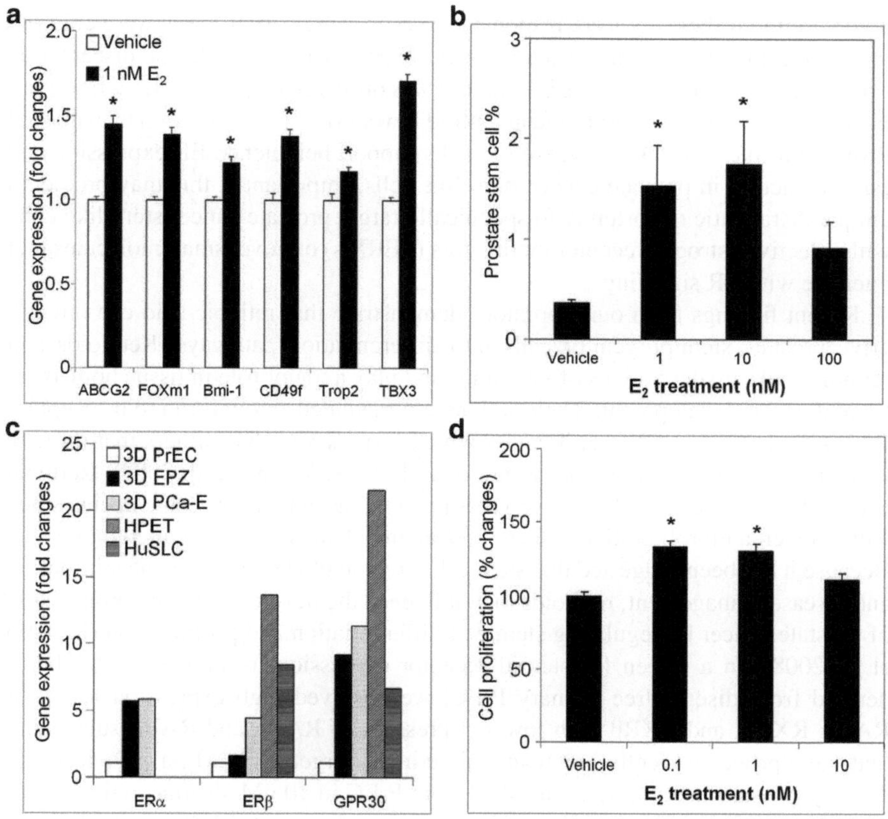


**Fig. 1.4** Modulation of prostate stem and progenitor cell populations by estrogen. (**a**) Prostaspheres were cultured from disease-free primary epithelial cells in the absence or presence of 1 nM of estradiol-17β ($E_2$) for 7 days, and gene expression was evaluated by qRT-PCR. $E_2$ significantly increased mRNA levels of stem cell markers ABCG2, FOXm1, Bmi-1, CD49f, Trop2, and TBX3. *$P<0.05$ vs. vehicle; $n=4$. (**b**) 2D primary prostate epithelial cells from disease-free donors were cultured for 72 h in 1, 10, or 100 nM $E_2$, and the Hoechst 33342 exclusion based side-population analysis by flow cytometry was used to measure the percentage of stem-like cells. *$P<0.05$ vs. vehicle by ANOVA; $n=6$. (**c**) ER expression by q RT-PCR in normal prostate stem-like cells and prostate cancer stem-like cells. A 3D prostasphere assay was used to isolate and amplify the stem/early-stage progenitor cell populations from disease-free primary prostate epithelial cultures (PrEC), matched prostate epithelial primary cultures from benign regions (EPZ) and prostate cancer cores with>80% cancer cells from the same patient (PCa-E). HPET and HuSLC are two human prostate cancer stem-like cell lines established from Gleason score 9 human prostate cancer (kindly supplied by Dr. Susan Kasper). Data is normalized to ER expression levels in PrEC-derived prostaspheres which was set as 1. (**d**) HuSLC cells were cultured in the absence or presence of 0.1–10 nM of $E_2$ for 72 h. Cell proliferation was evaluated by MTT assay. Treatment of 0.1 and 1 nM of $E_2$ significantly increased the HuSLC cell proliferation. *$P<0.05$ vs. vehicle; $n=4$

original tumors *in vivo*. While both prostate cancer stem-like cell lines were negative for ERα, they expressed 10–15-fold higher ERβ and 7–20-fold higher GPR30 levels compared to prostasphere cells from normal prostate epithelium (Fig. 1.4c).

To determine if these cells are responsive to estrogens, HuSLC cells were cultured for 72 h in increasing concentrations of $E_2$ (Fig. 1.4d). Similar to the disease-free PrEC primary cultures, a biphasic response was observed on proliferation; however, the stimulatory effects were found at tenfold lower doses (0.1–1 nM $E_2$) than normal PrEC stem-like cells. Together, these results support heightened ER expression and estrogen action in prostate cancer stem-like cells. Importantly, this may provide a unique therapeutic opportunity to specifically target prostate cancer stem-like cells with selective estrogen receptor modulators (SERMs) or novel small molecules that interfere with ER signaling.

Recent findings from our laboratory demonstrate that retinoic acid can directly drive prostate stem/progenitor cells into differentiation pathways. Retinoids and retinoic acids are derivatives of vitamin A and play a major role in tissue homeostasis and organ development. Their actions are mediated through retinoid receptors RARα, β, and γ and RXR α, β, and γ which form RAR/RXR dimers that directly regulate target gene expression (Metallo et al. 2008; Vezina et al. 2009). Retinoic acid and its synthetic analogs have great potential as anticarcinogenic agents since they trigger antiproliferative effects and augment differentiation in tumor cells. Because it has been suggested that stem cells are potential targets of cancer initiation and disease management, retinoids may influence the development and progression of prostate cancer by regulating stem cell differentiation and proliferation (Metallo et al. 2008). In a screen for steroid receptor expression in human prostaspheres derived from disease-free primary PrEC, we observed high expression levels of RARγ, RXRα, and RXRβ with lower expression of RARα and RARβ suggesting that stem/progenitor cells are potential retinoid targets (Fig. 1.5a). To test this directly, prostaspheres were established from PrEC in 10 nM all-trans retinoic acid (ATRA) and at day 4, spheroids were dispersed, stained with CD49f antibodies and dividing stem cells classified for division type (Fig. 1.1c). In vehicle cultures, 50% of stem cell divisions were symmetric self-renewal, and 18% were asymmetric cell divisions to generate a daughter progenitor cell (Fig. 1.5b). Exposure to ATRA shifted this to 28% symmetric stem cell self-renewal and 40% asymmetric cell division indicating that retinoic acid was driving the stem cells to enter a differentiation pathway. Continued culture of prostaspheres in ATRA resulted in advanced differentiation by day 7 as compared to vehicle cultures with increased size, bilayer formation, lumen initiation (Fig. 1.5c), and induction of AR, NKX3.1, and HOXB13 expression, markers of luminal epithelial cell differentiation (Fig. 1.5d). These findings support the multiple studies on the usefulness of retinoids in chemoprevention and treatment for prostate cancer (Huss et al. 2004; Schenk et al. 2009) and suggest that their actions may, in part, be mediated through direct actions on prostate stem cells, driving differentiation and limiting self-renewal. Thus, we predict that chemicals which either augment or interfere with retinoid signaling will have the capacity to directly alter human prostate stem cell differentiation with potentially beneficial or detrimental outcomes with regards to prostate health.

Several other members of the steroid receptor gene superfamily are highly expressed in the day 7 prostaspheres from disease-free PrEC including glucocorticoid receptor (GR), estrogen-related receptor-α (ERRα), aryl hydrocarbon receptor (AhR), peroxisome proliferator-activated receptor (PPAR)-γ, vitamin D receptor

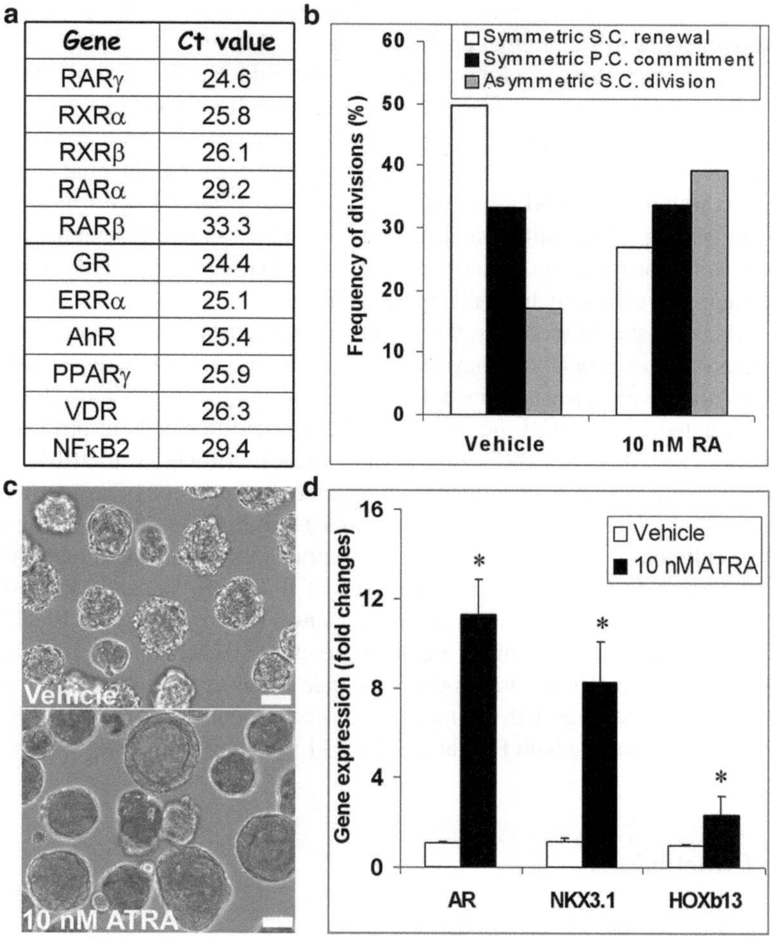


**Fig. 1.5** Prostate stem/progenitor cells differentiation by retinoic acid. (**a**). PCR Array analysis (SA Biosciences) of nuclear receptor superfamily gene expression in day 7 prostaspheres cultured from primary PrEC. Ct<35 cycles considered detectable expression. (**b**) Stem cell division types observed in day 4 prostasphere cells as determined by $CD49f^{high}$ staining (see Fig. 1.1c). Division frequency in vehicle-treated spheroids was ~50% symmetric stem cell renewal, ~32% symmetric division of committed progenitor cells, and ~18% cells asymmetric division to one stem cell and one daughter progenitor cell. Culture in 10 nM ATRA shifted this cell division pattern to 28% symmetric stem cell self-renewal and 40% asymmetric stem cell division. Total 100~150 pairs of daughter cells/group were counted. (**c**) Culture of prostaspheres to 7 days in basal media (*top*) or 10 nM ATRA (*bottom*). ATRA stimulated spheroid growth and formation of double-layered epithelial cells compared to vehicle. Bar=50 μm. (**d**) Culture of prostaspheres in 10 nM ATRA significantly increased AR, NKX3.1, and HOXB13 mRNA levels (qRT-PCR) compared to vehicle. $*P<0.05$. $n=4$

(VDR), and nuclear factor kappa beta (NFκB2) (Fig. 1.5a). These receptors and their cognate ligands may thus have potential roles in regulating either stem cells or progenitor cell populations in the prostate gland. In support of this, a recent study documented that 1,25-dihydroxyvitamin $D_3$ induced cell cycle arrest in mouse prostate epithelial stem/progenitor cell populations and drove them into differentiation pathways towards a luminal epithelial cell fate (Maund et al. 2011). Similar to retinoid treatment, these findings suggest a possible chemopreventive role for vitamin $D_3$ by targeting stem and progenitor cells to limit their proliferation and maintain epithelial differentiation phenotypes. Another recent study demonstrated that human prostate tumor-initiating cells with sphere-forming potential and multipotency exhibit increased NFκB signaling (Rajasekhar et al. 2011). Further, specific inhibitors of NFκB activation blocked their secondary sphere formation capacity and *in vivo* tumor initiation in animals suggesting that targeting NFκB in prostate cancer stem-like cells may have therapeutic efficacy.

In addition to steroid receptors, there is emerging evidence that prostate stem-like cells and progenitor populations may also be targets for protein hormones. Studies with transgenic mouse models determined that local prostate-specific prolactin (PRL) production resulted in a marked increase in the Sca-1 stem-like cells in prostate ducts which was associated with PRL-induced tumorigenesis (Rouet et al. 2010). Recent findings from our laboratory have also found significant expression of growth hormone (GH) receptor and insulin-like growth factor (IGF)-1 receptor in second passage prostaspheres cultured from Sprague–Dawley rat ventral, dorsal, and lateral lobes (unpublished studies with Dr. S. Swanson, University of Illinois at Chicago). Interestingly, these spheroids exhibited significant growth responses to exogenous GH (10–100 nM) suggesting that the stem/progenitors are direct targets of GH action in controlling prostate size and growth. These early findings indicate potential for regulation of normal and pathologic prostate growth via targeting this cell population with available pharmaceuticals that interfere with PRL and GH/IGF-1 signaling pathways.

## 1.5 Conclusions

In summary, although prostate epithelial stem cells and early-stage progenitor cells are AR negative and not considered direct androgen targets, there is emerging evidence to indicate that they are direct targets of multiple other steroid and protein hormones that can regulate their proliferative as well as differentiation status. This is schematized in Fig. 1.6 which proposes a hierarchical stem cell model (although this could equally apply to a bifurcated system) where some hormones including estrogens, GH, PRL, and others such as NFκB maintain stem/progenitor cell self-renewal and homeostasis whereas others including retinoids and vitamin $D_3$ stimulate the stem and progenitor populations to differentiate towards lineage cells. We propose that this tight balance of signals is involved in maintaining the stem cell niche and homeostasis within the prostate epithelium during normal development and through adulthood, responding to conditions as needed. It is possible that dysregulation of this balanced homeostasis contributes to prostate carcinogenesis and progression. A detailed insight into this regulatory system within the epithelial stem/progenitor cell populations may provide novel opportunities for chemoprevention and therapeutics for prostate cancer in the future.

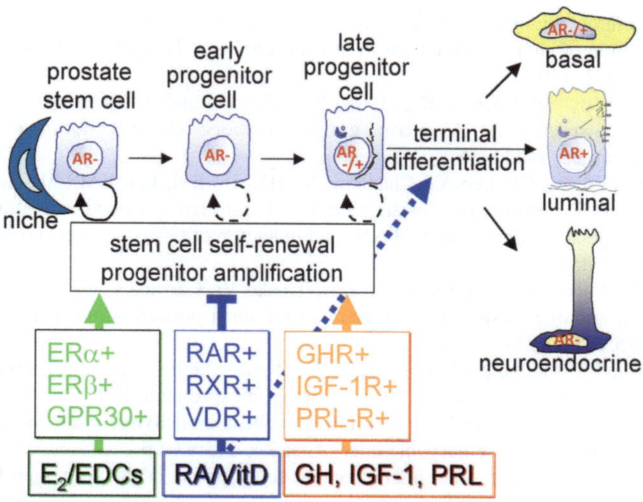


**Fig. 1.6** Hormones directly target prostate stem/progenitor cells. Prostate stem cell hierarchy shows self-renewing prostate stem cells give rise to transit-amplifying progenitor cells. These cells possess transient self-renewal activity and produce large numbers of terminally differentiated basal cells and secretory luminal cells as well as a small number of neuroendocrine cells. While androgen receptor is not detectable, prostate stem/early-stage progenitor cells do express robust levels of steroid receptors including ERs and GPR30, RARs/RXRs, VDR, PRL, and GHR/IGF-1R. When activated by their cognate ligands, these receptors mediate diverse effects including stem cell self-renewal, progenitor cell amplification, and entry into differentiated lineage pathways. Together, these hormones influence development and homeostasis of the normal prostate gland and may play important roles in prostate cancer initiation and progression

**Acknowledgements** The authors acknowledge the assistance of several individuals for their generous contributions towards data highlighted in this text: Dr. Susan Kasper (University of Cincinnati) and Dr. Larisa Nonn (University of Illinois at Chicago) for provisions of cells; Dr. Steve Swanson (University of Illinois at Chicago) for GHR and IGF-1R analysis; Dr. Guang-Bin Shi, Dan-Ping Hu, and Jacqueline Rinaldi for technical assistance; and Lynn Birch for editorial assistance. The research is supported by NIH grant RC2 ES018758.

## References

Asselin-Labat ML, Vaillant F, Sheridan JM, Pal B, Wu D, Simpson ER, Yasuda H, Smyth GK, Martin TJ, Lindeman GJ et al (2010) Control of mammary stem cell function by steroid hormone signalling. Nature 465:798–802

Berry PA, Maitland NJ, Collins AT (2008) Androgen receptor signalling in prostate: effects of stromal factors on normal and cancer stem cells. Mol Cell Endocrinol 288:30–37

Bhatia-Gaur R, Donjacour AA, Sciavolino PJ, Kim M, Desai N, Young P, Norton CR, Gridley T, Cardiff RD, Cunha GR et al (1999) Roles for Nkx3.1 in prostate development and cancer. Gen Devel 13:966–977

Bhatt RI, Brown MD, Hart CA, Gilmore P, Ramani VA, George NJ, Clarke NW (2003) Novel method for the isolation and characterisation of the putative prostatic stem cell. Cytometry A 54:89–99

Brown MD, Gilmore PE, Hart CA, Samuel JD, Ramani VA, George NJ, Clarke NW (2007) Characterization of benign and malignant prostate epithelial Hoechst 33342 side populations. Prostate 67:1384–1396
Chan KW, Yu KL, Rivier J, Chow BK (1998) Identification and characterization of a receptor from goldfish specific for a teleost growth hormone-releasing hormone-like peptide. Neuroendocrinology 68:44–56
Chan QK, Lam HM, Ng CF, Lee AY, Chan ES, Ng HK, Ho SM, Lau KM (2010) Activation of GPR30 inhibits the growth of prostate cancer cells through sustained activation of Erk1/2, c-jun/c-fos-dependent upregulation of p21, and induction of G(2) cell-cycle arrest. Cell Death Differ 17:1511–1523
Cocciadiferro L, Miceli V, Kang KS, Polito LM, Trosko JE, Carruba G (2009) Profiling cancer stem cells in androgen-responsive and refractory human prostate tumor cell lines. Ann NY Acad Sci 1155:257–262
Collins AT, Habib FK, Maitland NJ, Neal DE (2001) Identification and isolation of human prostate epithelial stem cells based on alpha(2)beta(1)-integrin expression. J Cell Sci 114:3865–3872
Cunha GR (1973) The role of androgens in the epithelio-mesenchymal interactions involved in prostatic morphogenesis in embryonic mice. Anat Rec 175:87–96
Cunha GR (1976) Epithelial-stromal interactions in development of the urogenital tract. Int Rev Cytol 47:137–194
Cunha GR (1984) Prostatic epithelial morphogenesis, growth, and secretory cytodifferentiation are elicited via trophic influences from mesenchyme. In: Progress in cancer research and therapy (ed. F. Bresciani), 31:121–128. New York: Raven Press
Cunha G, Fujii H, Neubauer B, Shannon J, Sawyer L, Reese B (1983) Epithelial-mesenchymal interactions in prostatic development. I. Morphological observations of prostatic induction by urogenital sinus mesenchyme in epithelium of the adult rodent urinary bladder. J Cell Biol 96:1662–1670
Cunha GR, Donjacour AA, Cooke PS, Mee S, Bigsby RM, Higgins SJ, Sugimura Y (1987) The endocrinology and developmental biology of the prostate. Endocr Rev 8:338–363
Dagvadorj A, Collins S, Jomain JB, Abdulghani J, Karras J, Zellweger T, Li H, Nurmi M, Alanen K, Mirtti T et al (2007) Autocrine prolactin promotes prostate cancer cell growth via Janus kinase-2-signal transducer and activator of transcription-5a/b signaling pathway. Endocrinology 148:3089–3101
Ding XW, Wu JH, Jiang CP (2010) ABCG2: a potential marker of stem cells and novel target in stem cell and cancer therapy. Life Sci 86:631–637
Donjacour AA, Cunha GR (1993) Assessment of prostatic protein secretion in tissue recombinants made of urogenital sinus mesenchyme and urothelium from normal or androgen-insensitive mice. Endocrinology 132:2342–2350
Donjacour AA, Cunha GR (1995) Induction of prostatic morphology and secretion in urothelium by seminal vesicle mesenchyme. Development 121:2199–2207
Garraway LA, Lin D, Signoretti S, Waltregny D, Dilks J, Bhattacharya N, Loda M (2003) Intermediate basal cells of the prostate: in vitro and in vivo characterization. Prostate 55:206–218
Garraway IP, Sun W, Tran CP, Perner S, Zhang B, Goldstein AS, Hahm SA, Haider M, Head CS, Reiter RE et al (2010) Human prostate sphere-forming cells represent a subset of basal epithelial cells capable of glandular regeneration in vivo. Prostate 70:491–501
Goldstein AS, Lawson DA, Cheng D, Sun W, Garraway IP, Witte ON (2008) Trop2 identifies a subpopulation of murine and human prostate basal cells with stem cell characteristics. Proc Natl Acad Sci USA 105:20882–20887
Goldstein AS, Huang J, Guo C, Garraway IP, Witte ON (2010a) Identification of a cell of origin for human prostate cancer. Science 329:568–571
Goldstein AS, Stoyanova T, Witte ON (2010b) Primitive origins of prostate cancer: In vivo evidence for prostate-regenerating cells and prostate cancer-initiating cells. Mol Oncol 4:385–396
Goodell MA, Brose K, Paradis G, Conner AS, Mulligan RC (1996) Isolation and functional properties of murine hematopoietic stem cells that are replicating in vivo. J Exp Med 183:1797–1806

Goto K, Salm SN, Coetzee S, Xiong X, Burger PE, Shapiro E, Lepor H, Moscatelli D, Wilson EL (2006) Proximal prostatic stem cells are programmed to regenerate a proximal-distal ductal axis. Stem Cells 24:1859–1868

Gu G, Yuan J, Wills M, Kasper S (2007) Prostate cancer cells with stem cell characteristics reconstitute the original human tumor in vivo. Cancer Res 67:4708–4715

Guo C, Liu H, Zhang BH, Cadaneanu RM, Mayle AM, Garraway IP (2012) Epcam, CD44, and CD49f distinguish sphere-forming human prostate basal cells from a subpopulation with predominant tubule initiation capability. PLoS One 7:e34219

Hayashi N, Sugimura Y, Kawamura J, Donjacour AA, Cunha GR (1991) Morphological and functional heterogeneity in the rat prostatic gland. Biol Reprod 45:308–321

Hayward S, Cunha G, Dahiya R (1996) Normal development and carcinogenesis of the prostate; a unifying hypothesis. Ann NY Acad Sci 784:50–62

Hu WY, Shi GB, Lam HM, Hu DP, Ho SM, Madueke IC, Kajdacsy-Balla A, Prins GS (2011) Estrogen-initiated transformation of prostate epithelium derived from normal human prostate stem-progenitor cells. Endocrinology 152:2150–2163

Hu WY, Shi GB, Hu DP, Nelles JL, Prins GS (2012) Actions of estrogens and endocrine disrupting chemicals on human prostate stem/progenitor cells and prostate cancer risk. Mol Cell Endocrinol 354:63–73

Huang L, Pu Y, Hu WY, Birch L, Luccio-Camelo D, Yamaguchi T, Prins GS (2009) The role of Wnt5a in prostate gland development. Dev Biol 328(2):188– 199

Hudson DL (2004) Epithelial stem cells in human prostate growth and disease. Prostate Cancer Prostatic Dis 7:188–194

Hudson DL, O'Hare M, Watt FM, Masters JR (2000) Proliferative heterogeneity in the human prostate: evidence for epithelial stem cells. Lab Invest 80:1243–1250

Huss JJ, Lai L, Barrios RJ, Hirschi KK, Greenberg NM (2004) Retinoic acid slows progression and promotes apoptosis of spontaneous prostate cancer. Prostate 61:142–152

Isaacs JT (2008) Prostate stem cells and benign prostatic hyperplasia. Prostate 68:1025–1034

Isaacs JT, Coffey DS (1989) Etiology and disease process of benign prostatic hyperplasia. Prostate Suppl 2:33–50

Isaacs JT, Barrack ER, Isaacs WB, Coffey DS (1981) The relationship of cellular structure and function: the matrix system. Prog Clin Biol Res 75A:1–24

Joshi PA, Jackson HW, Beristain AG, Di Grappa MA, Mote PA, Clarke CL, Stingl J, Waterhouse PD, Khokha R (2010) Progesterone induces adult mammary stem cell expansion. Nature 465: 803–807

Kasper S (2007) Characterizing the prostate stem cell. J Urol 178:375

Kasper S (2008) Exploring the origins of the normal prostate and prostate cancer stem cell. Stem Cell Rev 4:193–201

Kasper S (2009) Identification, characterization, and biological relevance of prostate cancer stem cells from clinical specimens. Urol Oncol 27:301–303

Lawson DA, Zong Y, Memarzadeh S, Xin L, Huang J, Witte ON (2010) Basal epithelial stem cells are efficient targets for prostate cancer initiation. Proc Natl Acad Sci USA 107:2610–2615

Leong KG, Wang BE, Johnson L, Gao WQ (2008) Generation of a prostate from a single cell. Nature 456:804–808

Liu AY, True L (2002) Characterization of prostate cell types by CD cell surface molecules. Am J Pathol 160:37–43

Liu AY, Roudier MP, True LD (2004) Heterogeneity in primary and metastatic prostate cancer as defined by cell surface CD profile. Am J Pathol 165:1543–1556

Liu J, Pascal LE, Isharwal S, Metzger D, Ramos Garcia R, Pilch J, Kasper S, Williams K, Basse PH, Nelson JB et al (2011) Regenerated luminal epithelial cells are derived from preexisting luminal epithelial cells in adult mouse prostate. Mol Endocrinol 25:1849–1857

Long RM, Morrissey C, Fitzpatrick JM, Watson RW (2005) Prostate epithelial cell differentiation and its relevance to the understanding of prostate cancer therapies. Clin Sci (Lond) 108:1–11

Lowsley OS (1912) The development of the human prostate gland with reference to the development of other structures at the neck of the urinary bladder. Am J Anat 13:299–348

Lukacs RU, Goldstein AS, Lawson DA, Cheng D, Witte ON (2010) Isolation, cultivation and characterization of adult murine prostate stem cells. Nat Protoc 5:702–713

Maitland NJ, Collins AT (2008) Prostate cancer stem cells: a new target for therapy. J Clin Oncol 26:2862–2870

Maitland NJ, Frame FM, Polson ES, Lewis JL, Collins AT (2011) Prostate cancer stem cells: do they have a basal or luminal phenotype? Horm Cancer 2:47–61

Mathew G, Timm EA Jr, Sotomayor P, Godoy A, Montecinos VP, Smith GJ, Huss WJ (2009) ABCG2-mediated DyeCycle Violet efflux defined side population in benign and malignant prostate. Cell Cycle 8:1053–1061

Maund SL, Barclay WW, Hover LD, Axanova LS, Sui G, Hipp JD, Fleet JC, Thorburn A, Cramer SD (2011) Interleukin-1alpha mediates the antiproliferative effects of 1,25-dihydroxyvitamin D3 in prostate progenitor/stem cells. Cancer Res 71:5276–5286

McCormick DL, Rao KVN, Steele VE, Lubet RA, Kelloff GJ, Bosland MC (1999) Chemoprevention of rat prostate carcinogenesis by 9-cis-retinoic acid. Canc Res 59:521–524

McNeal JE (1983) The prostate gland: morphology and pathobiology. Monogr Urol 4:3–11

McPherson SJ, Hussain S, Balanathan P, Hedwards SL, Niranjan B, Grant M, Chandrasiri UP, Toivanen R, Wang Y, Taylor RA et al (2010) Estrogen receptor-beta activated apoptosis in benign hyperplasia and cancer of the prostate is androgen independent and TNFalpha mediated. Proc Natl Acad Sci USA 107:3123–3128

Metallo CM, Ji L, de Pablo JJ, Palecek SP (2008) Retinoic acid and bone morphogenetic protein signaling synergize to efficiently direct epithelial differentiation of human embryonic stem cells. Stem Cells 26:372–380

Miki J, Rhim J (2008) Prostate cell cultures as in vitro models for the study of normal stem cells and cancer stem cells. Prostate Cancer Prostatic Dis 11:32–39

Moore KL, Persaud TVN (2008) Before we are born, essentials of embryology and birth defects, 7th edn. Saunders/Elsevier, Philadelphia

Morrison SJ, Kimble J (2006) Asymmetric and symmetric stem-cell divisions in development and cancer. Nature 441:1068–1074

Oldridge EE, Pellacani D, Collins AT, Maitland NJ (2012) Prostate cancer stem cells: are they androgen-responsive? Mol Cell Endocrinol 360(1–2):14–24

Presnell SC, Petersen B, Heidaran M (2002) Stem cells in adult tissues. Semin Cell Dev Biol 13:369–376

Prins GS (1993) Development of the prostate. In: Reproductive issues and the aging male (Eds. F. Haseltine, A. Paulsen, C. Wang), pp 101–112. Embryonic, Inc, New York

Prins GS, Birch L (1995) The developmental pattern of androgen receptor expression in rat prostate lobes is altered after neonatal exposure to estrogen. Endocrinology 136:1303–1314

Prins GS, Korach KS (2008) The role of estrogens and estrogen receptors in normal prostate growth and disease. Steroids 73:233–244

Prins GS, Putz O (2008) Molecular signaling pathways that regulate prostate gland development. Differentiation 6:641–659

Prins GS, Cooke PS, Birch L, Donjacour AA, Yalcinkaya TM, Siiteri PK, Cunha GR (1992) Androgen receptor expression and 5a-reductase activity along the proximal-distal axis of the rat prostatic duct. Endocrinology 130:3066–3073

Prins GS, Birch L, Woodham C (1995) Effect of neonatal estrogen on androgen receptor protein and mRNA levels in developing rat ventral prostate. Paper presented at: 77th annual meeting of the endocrine society, Washington, DC

Rajasekhar VK, Studer L, Gerald W, Socci ND, Scher HI (2011) Tumour-initiating stem-like cells in human prostate cancer exhibit increased NF-kappaB signalling. Nat Commun 2:162

Richardson GD, Robson CN, Lang SH, Neal DE, Maitland NJ, Collins AT (2003) CD133, a novel marker for human prostatic epithelial stem cells. J Cell Sci 117:3539–3545

Ricke WA, McPherson SJ, Bianco JJ, Cunha GR, Wang Y, Risbridger GP (2008) Prostatic hormonal carcinogenesis is mediated by in situ estrogen production and estrogen receptor alpha signaling. FASEB J 22:1512–1520

Robinson EJ, Neal DE, Collins AT (1998) Basal cells are progenitors of luminal cells in primary cultures of differentiating human prostatic epithelium. Prostate 37:149–160

Rouet V, Bogorad RL, Kayser C, Kessal K, Genestie C, Bardier A, Grattan DR, Kelder B, Kopchick JJ, Kelly PA et al (2010) Local prolactin is a target to prevent expansion of basal/stem cells in prostate tumors. Proc Natl Acad Sci USA 107:15199–15204

Rumpold H, Heinrich E, Untergasser G, Hermann M, Pfister G, Plas E, Berger P (2002) Neuroendocrine differentiation of human prostatic primary epithelial cells in vitro. Prostate 53:101–108

Scaffidi P, Misteli T (2011) In vitro generation of human cells with cancer stem cell properties. Nat Cell Biol 13:1051–1061

Schalken JA (2007) Prostate cancer stem cells. In: Prostate cancer: biology, genetics and the new therapeutics (Eds. L. Chung, W. Isaacs, J. Simons), pp 63-72, 2nd edn. Humana, Totowa

Schenk JM, Riboli E, Chatterjee N, Leitzmann MF, Ahn J, Albanes D, Reding DJ, Wang Y, Friesen MD, Hayes RB et al (2009) Serum retinol and prostate cancer risk: a nested case-control study in the prostate, lung, colorectal, and ovarian cancer screening trial. Cancer Epidemiol Biomarkers Prev 18:1227–1231

Sharifi N, Hurt EM, Farrar WL (2008) Androgen receptor expression in prostate cancer stem cells: is there a conundrum? Cancer Chemother Pharmacol 62:921–923

Sissung TM, Danesi R, Kirkland CT, Baum CE, Ockers SB, Stein EV, Venzon D, Price DK, Fig. WD (2011) Estrogen receptor a and aromatase polymorphisms affect risk, prognosis, and therapeutic outcome in men with castration-resistant prostate cancer treated with docetaxel-based therapy. J Clin Endocrinol Metab 95:E368–E372

Smith S, Neaves W, Teitelbaum S (2007) Adult versus embryonic stem cells: treatments. Science 316:1422–1423

Timms BG, Mohs TJ, Didio LJ (1994) Ductal budding and branching patterns in the developing prostate. J Urol 151:1427–1432

Tomasetti C, Levy D (2010) Role of symmetric and asymmetric division of stem cells in developing drug resistance. Proc Natl Acad Sci USA 107:16766–16771

Tsujimura A, Koikawa Y, Salm S, Takao T, Coetzee S, Moscatelli D, Shapiro E, Lepor H, Sun TT, Wilson EL (2002) Proximal location of mouse prostate epithelial stem cells: a model of prostatic homeostasis. J Cell Biol 157:1257–1265

Vander Griend DJ, Karthaus WL, Dalrymple S, Meeker A, DeMarzo AM, Isaacs JT (2008) The role of CD133 in normal human prostate stem cells and malignant cancer-initiating cells. Cancer Res 68:9703–9711

Vander Griend DJ, D'Antonio J, Gurel B, Antony L, Demarzo AM, Isaacs JT (2010) Cell-autonomous intracellular androgen receptor signaling drives the growth of human prostate cancer initiating cells. Prostate 70:90–99

Vermeulen A, Kaufman JM, Goemaere S, van Pottelberg I (2002) Estradiol in elderly men. Aging Male 5:98–102

Vezina CM, Lin TM, Peterson RE (2009) AHR signaling in prostate growth, morphogenesis, and disease. Biochem Pharmacol 77:566–576

Wang ZA, Shen MM (2011) Revisiting the concept of cancer stem cells in prostate cancer. Oncogene 30:1261–1271

Wang Y, Hayward S, Cao M, Thayer K, Cunha G (2001a) Cell differentiation lineage in the prostate. Differentiation 68:270–279

Wang Y, Chang WY, Prins GS, van Breeman RB (2001b) Simultaneous determination of all-trans, 9-cis, 13-cis retinoic acid and retinal in rat prostate using liquid. J Mass Spectrom 36:882–888

Wang Z, Prins GS, Coschigano KT, Kopchick JJ, Green JE, Ray VH, Hedayat S, Christov KT, Unterman TG, Swanson SM (2005) Disruption of growth hormone signaling retards early stages of prostate carcinogenesis in the C3(1)/T antigen mouse. Endocrinology 146:5188–5196

Wang G, Wang Z, Sarkar FH, Wei W (2012) Targeting prostate cancer stem cells for cancer therapy. Discov Med 13:135–142

Wu M, Kwon HY, Rattis F, Blum J, Zhao C, Ashkenazi R, Jackson TL, Gaiano N, Oliver T, Reya T (2007) Imaging hematopoietic precursor division in real time. Cell Stem Cell 1:541–554

Xin L, Lawson DA, Witte ON (2005) The Sca-1 cell surface marker enriches for a prostate-regenerating cell subpopulation that can initiate prostate tumorigenesis. Proc Natl Acad Sci USA 102:6942–6947

Xin L, Lukacs RU, Lawson DA, Cheng D, Witte ON (2007) Self-renewal and multilineage differentiation in vitro from murine prostate stem cells. Stem Cells 25:2760–2769

Zhou S, Schuetz JD, Bunting KD, Colapietro AM, Sampath J, Morris JJ, Lagutina I, Grosveld GC, Osawa M, Nakauchi H et al (2001) The ABC transporter Bcrp1/ABCG2 is expressed in a wide variety of stem cells and is a molecular determinant of the side-population phenotype. Nat Med 7:1028–1034

# Chapter 2
# Isolation and Characterization of Prostate Stem Cells

**Andrew S. Goldstein and Owen N. Witte**

**Abstract** Based on the unique capacity of the rodent prostate to undergo seemingly endless rounds of androgen cycling in response to castration and androgen add-back, the prostate has been proposed to contain long-term self-renewing stem cells. However the prospective isolation and characterization of stem-like cells from rodent and human prostate tissue has only been described over the last 2 decades. Several models of epithelial homeostasis in the adult prostate have been proposed based on either the presence of a multipotent tissue stem cell that differentiates through a series of intermediate developmental stages or the coexistence of

A.S. Goldstein, Ph.D. (✉)
Department of Molecular and Medical Pharmacology, David Geffen School of Medicine, University of California, Los Angeles, CA 90095, USA

Department of Urology, David Geffen School of Medicine, University of California, Los Angeles, CA 90095, USA

Jonsson Comprehensive Cancer Center, David Geffen School of Medicine, University of California, Los Angeles, CA 90095, USA

Eli and Edythe Broad Center of Regenerative Medicine and Stem Cell Research, University of California, Los Angeles, CA 90095, USA
e-mail: AGoldstein@mednet.ucla.edu

O.N. Witte, M.D.
Department of Molecular and Medical Pharmacology, David Geffen School of Medicine, University of California, Los Angeles, CA 90095, USA

Department of Microbiology, Immunology and Molecular Genetics, University of California, Los Angeles, CA 90095, USA

Jonsson Comprehensive Cancer Center, David Geffen School of Medicine, University of California, Los Angeles, CA 90095, USA

Eli and Edythe Broad Center of Regenerative Medicine and Stem Cell Research, University of California, Los Angeles, CA 90095, USA

S.D. Cramer (ed.), *Stem Cells and Prostate Cancer*, DOI 10.1007/978-1-4614-6498-3_2, 

multiple unipotent lineage-restricted stem cells. The isolation of cells with stem and progenitor activity is an important first step to delineate the epithelial hierarchy of the prostate. In addition, isolation of stem cells allows characterization of their functional capacities and the molecular programs regulating their activity. These studies will enable detection or targeting of stem and progenitor cells during various stages of neoplastic transformation and tumor progression, including the lethal phase of the disease, castration-resistant prostate cancer.

## 2.1 Introduction

While the existence of stem cells in the prostate has long been postulated, their isolation or purification for functional testing has only been described in the last decade. Pioneering androgen cycling experiments by Isaacs and colleagues (Isaacs 1985), later repeated by Wilson and colleagues (Tsujimura et al. 2002), show that long-term castration-resistant stem cells in the rodent prostate are capable of almost indefinitely regenerating the gland after castration-mediated prostate involution and androgen add-back. The adult prostate gland predominantly comprises basal cells and luminal secretory cells with a very minor component of neuroendocrine cells (Abate-Shen and Shen 2000; Shen and Abate-Shen 2010). While basal cells in the rodent prostate predominantly remain after castration, a subset of luminal cells also survive and may participate in gland regeneration upon administration of androgen (English et al. 1987). Divergent models have been proposed to describe prostate homeostasis, from a bipotent prostate stem cell capable of regenerating all mature cell types in the gland to the coexistence of multiple unipotent or lineage-restricted stem cells (Lawson and Witte 2007; Shen et al. 2008). Based on differential keratin stains in both the mouse and human prostate, some have proposed a developmental model starting from a primitive stem cell that undergoes maturation and differentiation through several intermediate cell stages (Okada et al. 1992; Xue et al. 1998; van Leenders et al. 2000, 2003; Hudson 2004). However, functional evidence of such a hierarchical structure has been lacking until recently.

In order to accurately define the prostate epithelial hierarchy based on functional studies, isolation of prostate stem and progenitor cells is an essential first step. In addition to delineating hierarchical relationships, stem cell isolation enables both molecular and functional characterization to determine the capacity of isolated stem cells and define the pathways regulating stem-like behavior. A comprehensive analysis of stem cell properties may lead to targeted therapies of stem-like cells in various stages of disease including the lethal castration-resistant prostate cancer (Feldman and Feldman 2001; Shen and Abate-Shen 2010).

## 2.2 Methods of Stem Cell Isolation

A number of indirect approaches have been taken to investigate stem cells in the prostate, including the use of BrdU labeling to identify slow-cycling label-retaining cells (Tsujimura et al. 2002). While label retention is a property that some believe to be associated with stem cells (Cotsarelis et al. 1989; Berardi et al. 1995; Thorgeirsson 1996; Beauchamp et al. 2000; Lavker and Sun 2000; Slack 2000), this method alone does not provide functional evidence for stem cell activity. Importantly, intestinal tissue stem cells and hair follicle stem cells marked by the Wnt target gene Lgr5 are rapidly dividing (Barker et al. 2007; Jaks et al. 2008), indicating that quiescence is not a universal property of adult stem cells. In this chapter, we will describe the two general approaches that have been described to functionally define stem cell populations in the prostate: isolation of enriched cell preparations from dissociated mouse and human prostate tissues and lineage tracing in genetically engineered mouse models.

In the first approach, cells are purified and isolated from preparations of dissociated prostate tissue. Isolated cells, separated or fractionated into populations enriched for stem cells and those depleted for stem cells, are placed into various in vitro and in vivo functional assays to determine their inherent proliferative, clonogenic, and regenerative capacities. This strategy enables parallel identification of cell populations from naïve rodent (primarily mouse) and human tissues taken from surgical specimens (Lukacs et al. 2010a; Goldstein et al. 2011). Importantly, cell preparations are easily collected for analysis of DNA, RNA, or protein to determine what molecular characteristics distinguish stem cells from their non-stem cell counterparts. An alternative strategy described to define prostate stem cells is lineage tracing in genetically engineered mice. In this approach, genetic marking is performed in a stem cell leading to expression of a reporter protein in the stem cell and all of its progeny (Wang et al. 2009; Choi et al. 2012). This strategy (described in greater detail in Sect. 2.2.2.2) allows for demonstration of hierarchical relationships within the intact gland, maintaining important interactions with neighboring epithelial cells, non-epithelial stromal and immune cell populations, and other components of the local microenvironment.

### *2.2.1 In Vitro Assays to Measure Function of Isolated Stem/Progenitor Cells*

When taken out of their native site and grown in tissue culture, primitive cells should possess the capacity to extensively proliferate and self-renew under the appropriate conditions including growth factors and adhesive substratum (Barrandon and Green 1987; Ogawa 1993; Hudson et al. 2000; Uzgare et al. 2004; Lukacs et al. 2008). In contrast, more mature or differentiated cell populations, particularly those that are postmitotic, would be less likely to grow or persist long term. For this reason, clonogenic assays have been particularly useful in identifying cell subsets enriched for long-term immature cell activity (Ploemacher and Brons 1989; Reynolds and Weiss 1996; Dontu et al. 2003; Shackleton et al. 2006; Lim et al. 2009; Rock et al. 2009). Isolated cell populations can be tested using clonogenic in vitro assays that provide a quantitative measure of their proliferative and self-renewal activity. The two most commonly used assays are the colony-forming assay, measuring clonogenic and proliferative potential, and the sphere-forming assay, measuring both clonogenicity and self-renewal in vitro in a quantitative fashion (Fig. 2.1).

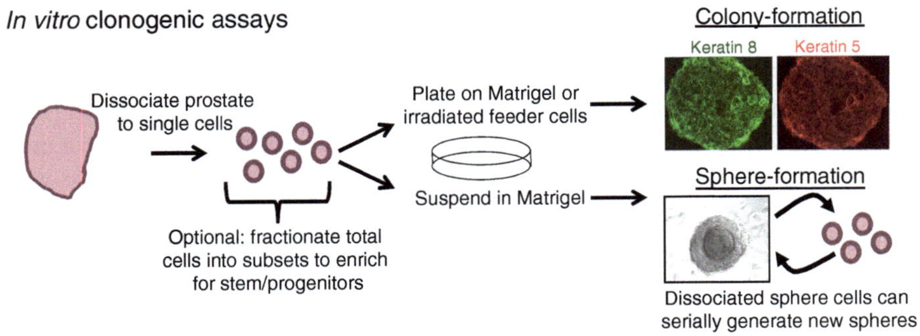


**Fig. 2.1** In vitro clonogenic assays. Mouse or human prostate tissues are dissociated to single cells. Subsets of epithelium can be fractionated from dissociated preparations to compare functional activity of enriched or depleted cell populations. Cells that are plated in a two-dimensional environment on irradiated feeder cells or Matrigel will form colonies, the majority of which express both luminal (keratin 8) and basal (keratin 5) epithelial keratins. Alternatively, cells that are suspended in a three-dimensional environment of Matrigel will form spheres that can be dissociated to single cells and serially replated to measure self-renewal activity

#### 2.2.1.1 Colony-Forming Assay

In the colony-forming assay, isolated cells are plated either directly onto a tissue culture dish, cocultured with irradiated feeder cells (such as adult fibroblasts, 3T3 fibroblasts, mouse embryonic fibroblasts), or on a matrix substratum (collagen or Matrigel) (Collins et al. 2001; Lawson et al. 2007; Lukacs et al. 2010a). Interestingly, colonies from mouse prostate co-express both basal (K5) and luminal (K8) keratins, representing an intermediate or putative transit-amplifying phenotype rarely observed in normal glands. When dissociated to single cells and serially replated, colonies exhibit limited self-renewal activity which is at least partially due to a Rho-kinase (ROCK)-mediated response, as inhibition of ROCK promotes colony self-renewal in vitro (Zhang et al. 2011). Alternatively, cultures of human prostate epithelial cells (PrECs) grown in low-calcium conditions demonstrate stem cell-like colony-forming activity (Litvinov et al. 2006).

While primary cells that form colonies maintain expression of some markers of cells in the gland, such as epithelial keratin expression, these cells lose the glandular structure characteristic of their native environment. Tsujimura et al. (2002) described a clonogenic assay where primary mouse prostate cells are suspended in collagen and grown in vitro to form ductal structures (Tsujimura et al. 2002). Cells from the proximal region are enriched for this activity and can generate glands containing distinct keratin 14+ basal and keratin 8+ luminal cells. Although a range of conditions can be used to measure colony-forming activity of naïve primary mouse and human prostate cells, each assay has been effectively utilized to compare the growth of distinct epithelial subpopulations (described further in Sect. 2.3). Recently, methods have been developed for long-term culture of mouse prostate stem cells that retain multi-lineage differentiation and self-renewal in vitro and in vivo (Barclay et al. 2008). These methods are discussed in more detail elsewhere in this book (Chaps. 9 and 10).

#### 2.2.1.2 Sphere-Forming Assay

To quantitatively measure self-renewal activity, primary cells grown within a three-dimensional matrix of Matrigel can be expanded in vitro (Lawson et al. 2007; Shi et al. 2007; Xin et al. 2007). Matrigel is rich in extracellular matrix proteins including laminin, collagen, and fibronectin (Emonard et al. 1987a, b) which are found in the basement membrane structures surrounding benign prostate glands in their native microenvironment (Bonkhoff et al. 1991; Fong et al. 1991). Dissociated single cells can be isolated from primary spheres and replated in Matrigel to generate secondary spheres, demonstrating self-renewal activity (Lawson et al. 2007; Xin et al. 2007). This activity can be repeated numerous times during serial replating or passaging to demonstrate the presence of long-term self-renewing cells (Lawson et al. 2007; Shi et al. 2007; Xin et al. 2007; Goldstein et al. 2008; Garraway et al. 2010).

The outer layer of cells in mouse prostate spheres comprises p63+ cells, analogous to the outer layer of p63+ basal cells in the gland, that are proliferating based on positive stains for Ki67. Proliferating p63+ cells appear to spontaneously differentiate toward the center or luminal space, which is filled with p63- cells undergoing apoptosis as marked by TUNEL staining (Xin et al. 2007). Sphere cells of both mouse and human origin retain in vivo stem-like activity to reconstitute glandular structures containing both basal and luminal epithelial cells when transplanted into mice (Shi et al. 2007; Xin et al. 2007; Garraway et al. 2009). Both spheres and colonies are clonally derived indicating that they arise from a single stem or progenitor cell (Lawson et al. 2007; Garraway et al. 2010). These assays allow for identification of markers to enrich for functional stem/progenitor cell subsets that can grow in vitro. They also provide quantitative measures of progenitor function that can be used to determine pathways and factors regulating stem and progenitor cell activity (Mulholland et al. 2009; Lukacs et al. 2010b; Shahi et al. 2011).

### 2.2.2 *In Vivo Assays to Measure Prostate Stem Cells*

While in vitro assays have been utilized to quantify the functional activity of isolated cells from adult epithelial tissues, colony and sphere-forming assays are believed to measure progenitor activity rather than true stem cell function. For example, far greater numbers of mammary stem cell-enriched $Lin^{-}CD24^{+}CD29^{hi}$ cells can generate clonogenic colonies in vitro than can repopulate mammary gland structures in vivo (Asselin-Labat et al. 2010). Therefore, in vivo assays are the most stringent tests of stem cell activity (Fig. 2.2), three of which will be described in the following section.

#### 2.2.2.1 In Vivo Tissue-Regeneration Assays to Measure Prospectively Isolated Stem Cell Populations

Cunha and colleagues first described a tissue fragment recombination assay where mid-gestation urogenital sinus, the region destined to develop into the prostate, can

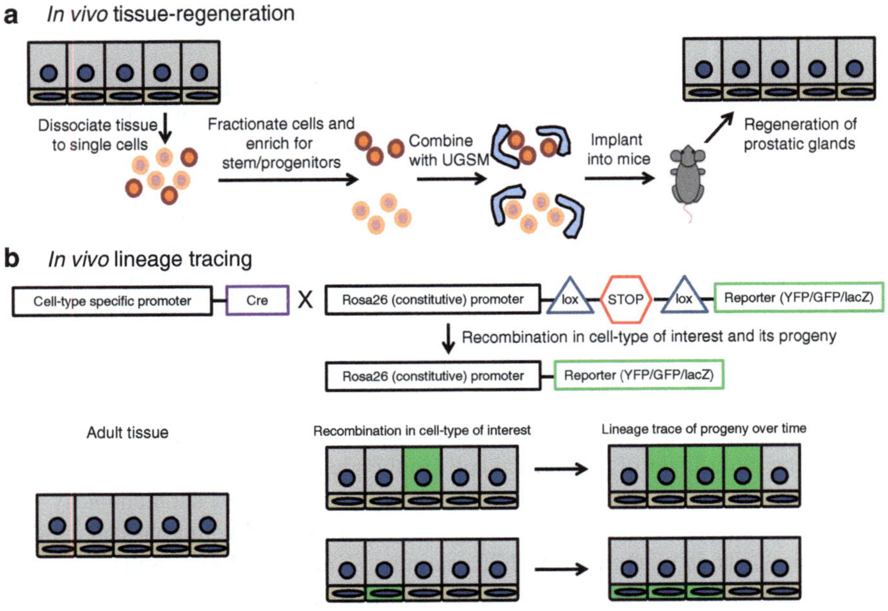


**Fig. 2.2** In vivo stem cell assays. (**a**) In the in vivo tissue-regeneration assay, prostate glands are dissociated to single cells and enriched/depleted subsets of cells are fractionated, combined with Urogenital Sinus Mesenchyme (UGSM), and implanted back into immune-deficient mice, either under the renal capsule or skin. Over 6–12 weeks, dissociated cells will regenerate prostatic glands. (**b**) An alternative approach is in vivo lineage tracing. The Cre recombinase, expressed in a cell-type-specific manner, will remove a STOP codon leading to constitutive expression of a reporter protein in the cell type of interest and all of its progeny

be physically separated into urogenital sinus mesenchyme (UGSM) and urogenital sinus epithelium (UGSE) and put back together under the renal capsule of immune-deficient mice (Cunha and Lung 1978). After a period of weeks to months, the recombined tissue will develop into prostatic glands. Our laboratory has made two important adaptations to utilize this assay with dissociated adult cells (Xin et al. 2003). First, we found that UGSM could support the regeneration of dissociated adult mouse prostate cells. The adult prostate could regenerate glands that resembled the native prostate gland (Xin et al. 2003). Since the epithelium was coming from the adult prostate, a second important adaptation was made to allow greater experimental control and perform transplantations without carefully timed matings. Xin et al. (2003) showed that UGSM could be dissociated to single cells, expanded in culture, frozen and thawed, and maintain its inductive activity on adult epithelium (Xin et al. 2003).

Since total dissociated adult mouse prostate cells could regenerate glands, the pool must contain stem-like tissue-regenerating cells, which could then be isolated using various approaches. Mouse prostate glandular regeneration is an extremely robust assay, used by many labs (Xin et al. 2003; Burger et al. 2005; Shi et al. 2007; Leong et al. 2008; Wang et al. 2009). A parallel dissociated cell tissue-regeneration assay using freshly isolated naïve benign primary human epithelial cells is more difficult for two reasons. First, obtaining fresh human tissue requires a considerable degree of effort and coordination between surgeons, pathologists, and researchers in a timely manner (Goldstein et al. 2011). Secondly, human prostate tissue is most commonly isolated from aged men who suffer from a disorder either in the prostate or in a neighboring tissue, such as the bladder. In contrast, studies with mouse tissue are generally performed on young, healthy tissue. Despite these difficulties, we demonstrated the use of an analogous tissue-regeneration assay where naïve human epithelial cells could regenerate human prostate glands in vivo (Goldstein et al. 2010, 2011).

It is important to note that while the tissue-recombination assay can be easily adapted based on cell type and species of origin, the approach requires taking adult cells out of their niche in the prostate gland. The process of gland dissociation could disrupt critical cell–cell interactions, remove signals from supportive cell populations, and preferentially select for one cell type over another in transplanted preparations. As mentioned in the introduction, an alternative approach to retain the native structure is genetic lineage tracing.

#### 2.2.2.2 In Vivo Lineage Tracing of Prostate Stem/Progenitor Cells and Their Progeny

The most common lineage tracing experiments utilize a cell-type or lineage-specific promoter to drive expression of the Cre recombinase either in a constitutive or regulated manner, reviewed by Fuchs and Horsley (2011). The Cre enzyme is a bacteriophage topoisomerase that specifically recognizes a short 34-base-pair stretch of DNA called a lox sequence, made up of two inverted repeats and a spacer region (Lakso et al. 1992). Mice engineered to express Cre from a lineage-specific promoter are most commonly bred with a reporter strain where lox sites flanking a STOP codon are placed between the ubiquitously expressed Rosa26 promoter and a reporter protein, such as green fluorescent protein (GFP) or lacZ (which can be detected upon the addition of the β-galactosidase substrate) (Soriano 1999; Barker et al. 2007; Fuchs and Horsley 2011). Using this strategy, the lox sites are excised allowing reporter expression only in the presence of the Cre recombinase. Since the recombination event occurs at the DNA level in the stem cell, all cells derived from that stem cell will retain the recombined allele and exhibit reporter expression. This strategy allows for robust tracing of the progeny from the original cells engineered to express the Cre protein. Therefore, the specificity and sensitivity of Cre expression in the cell type of interest are of vital importance for interpreting results.

#### 2.2.2.3 In Vivo Castration and Androgen-Mediated Regeneration to Demonstrate Stem Cell Activity

The most impressive display of prostate stem cell activity is in its capacity to survive androgen ablation or castration and promote regeneration of the gland upon administration of androgen (Tsujimura et al. 2002). The involution following castration and regeneration upon androgen add-back can be repeated almost indefinitely. Lukacs et al. (2008) utilized the tissue-recombination approach to demonstrate that adult prostate cells contain long-term self-renewing stem-like cells that can survive androgen deprivation and mediate regeneration following androgen add-back (Lukacs et al. 2008). Adult dissociated cells were combined with UGSM and implanted under the renal capsule of intact mice. Recipient mice were then subjected to rounds of castration-induced involution and androgen-mediated regeneration (Lukacs et al. 2008). In a manner analogous to the native prostate, regenerated tissue under the renal capsule was also capable of castration resistance and self-renewal. The lineage tracing approach has been combined with castration/regeneration to mark cells in the castrated or involuted state, using an inducible Cre recombinase to label cells with a fluorescent reporter protein, and then demonstrate the labeled progeny of those cells after androgen-mediated regeneration (Wang et al. 2009). These experiments prove that castration-resistant cells contribute to local tissue regeneration after addition of androgen.

## 2.3 Identification of Stem Cell Populations

### *2.3.1 Isolated Stem Cells in Dissociated Mouse Tissues*

Having developed assays to measure stem-like activity, numerous groups have now defined methods to purify cell populations enriched for stem/progenitor cells. Purification methods have been primarily based on cell-surface markers combined with Fluorescence-Activated Cell Sorting (FACS) although methods to enrich for functional or enzymatic activity have also been used for stem/progenitor purification. While the majority of studies from numerous laboratories implicate basal-like cells from the mouse prostate as stem/progenitor cells (Lawson et al. 2007, 2010; Goldstein et al. 2008; Burger et al. 2009; Choi et al. 2012), recent studies have also proven that the luminal epithelial layer contains stem cells (Wang et al. 2009; Choi et al. 2012).

#### 2.3.1.1 Evidence for Stem Cells with a Basal Localization

The stem cell antigen-1 (Sca-1), a marker of primitive stem/progenitor cells in many adult tissues (Spangrude et al. 1988; Welm et al. 2002; Kim et al. 2005), was used by two different groups to enrich for cells from the mouse prostate capable of tissue regeneration in vivo (Burger et al. 2005; Xin et al. 2005). While Sca-1+ cells, isolated from dissociated adult mouse prostate by FACS, can efficiently regenerate prostatic glands under the renal capsule, Sca-1- cells are devoid of this activity (Xin et al. 2005; Burger et al. 2009). A subset of Sca-1+ cells concentrated in the region most proximal to the urethra also express the Wnt target gene Axin2, indicating a potential role for Wnt signaling in maintaining these progenitor cells in their niche (Ontiveros et al. 2008). Sca-1 is expressed both in stromal and epithelial cells, indicating that additional markers are necessary to isolate epithelial stem cells (Lawson et al. 2010). Lawson et al. (2007) found that integrin alpha 6 (CD49f), which is a stem cell marker in several adult tissues (Stingl et al. 2006; Rock et al. 2009; Notta et al. 2011), could further enrich for cells in the mouse prostate capable of tissue regeneration in vivo (Lawson et al. 2007, 2010). Up to 1 in every 44 cells in this population exhibited colony formation in vitro on either irradiated 3T3 mouse fibroblast feeder cells (Lawson et al. 2007) or on Matrigel (Lawson et al. 2010). These stem cells were identified by depleting for lineage (Lin) antibodies against hematopoietic (CD45+), endothelial (CD31+), and red blood cells (Ter119) to generate a $Lin^{-}Sca\text{-}1^{+}CD49f^{hi}$ (LSC) profile. LSC cells identified a basally located stem cell population in the mouse prostate (Lawson et al. 2010).

The type I transmembrane protein Trop2, which is overexpressed in numerous cancers and associated with poor prognosis (Ohmachi et al. 2006; Fong et al. 2008a, b; Muhlmann et al. 2009; Kobayashi et al. 2010), was found to be highly expressed on a subset of mouse prostate LSC cells enriched for in vitro and in vivo stem-like activity (Goldstein et al. 2008). By gating on high levels of Trop2, up to 1/11 LSC $Trop2^{hi}$ cells from the mouse prostate could generate spheres in vitro (Goldstein et al. 2008). In vivo, DsRed labeled LSC $Trop2^{hi}$ cells could regenerate glands containing basal, luminal, and neuroendocrine cells. Importantly, neuroendocrine cells in regenerate glands were labeled with the DsRed protein, indicating that they were derived from donor stem cells capable of tri-lineage differentiation upon transplantation (Goldstein et al. 2008).

A single study implicates expression of the hematopoietic stem cell and germ cell marker ckit/CD117 (Manova et al. 1990; Ikuta and Weissman 1992) on putative mouse prostate epithelial stem cells and human prostate basal cells (Leong et al. 2008). However, numerous groups have found ckit expression in the adult human prostate to localize exclusively to the non-epithelial compartment on immune-infiltrating mast cells and specialized stromal cells termed interstitial cells of Cajal (ICC) (Van der Aa et al. 2003; Shafik et al. 2005; Nguyen et al. 2011). Other groups have reported an absence of ckit expression in mouse prostate luminal cells (Wang et al. 2009) or on adult mouse prostate epithelium (Blum et al. 2009).

While several surface markers can be combined to isolate basal stem cells from the mouse prostate, other approaches have also been utilized for mouse prostate stem cell purification. Cells with high levels of aldehyde dehydrogenase (ALDH) enzymatic activity can be labeled with a substrate that gets trapped inside of target cells and fluoresces at a detectable wavelength for FACS isolation. Burger et al. (2009) showed that $Aldefluor^{bright}$ cells are enriched for stem cell activity and that these cells predominantly localize to the basal cell layer (Burger et al. 2009). Finally, using a cyan fluorescent protein (CFP) reporter expressed from the basal cell-specific keratin 5 promoter, K5-CFP+ cells show stem cell properties when isolated and grown in vitro or in vivo (Peng et al. 2011).

## 2.3.2 *Identification of Stem Cells Using a Lineage Tracing Approach*

### 2.3.2.1 Luminal Stem Cells in the Castrated/Regressed Prostate

Using a genetic lineage tracing approach, Wang et al. (2009) identified a luminal stem cell population in the castrated/regressed mouse prostate (Wang et al. 2009). Although the androgen target gene and homeobox transcription factor Nkx3.1 is dramatically downregulated following castration, rare luminal cells in the regressed/involuted mouse prostate gland remain in Nkx3.1+. By engineering an inducible Nkx3.1 promoter to drive expression of the Cre allele, Shen and colleagues were able to label *CAstration-R*esistant Nkx3.1+ luminal cells (CARNs) and their progeny with YFP (Wang et al. 2009). After androgen add-back and regeneration of the gland, the authors found evidence of basal, luminal, and neuroendocrine cells labeled with YFP, indicating that CARNs represent a stem cell population capable of multi-lineage differentiation. CARNs were isolated in the castrated state and subjected to in vivo tissue regeneration to show that this stem cell population can also regenerate prostatic glands upon transplantation under the kidney capsule (Wang et al. 2009).

### 2.3.2.2 Parallel Identification of Unipotent Basal and Luminal Stem Cells in the Adult Mouse Prostate

Given the findings that in the normal prostate, basal cells are the predominant cell type capable of tissue regeneration upon transplantation, but rare luminal cells in the castrated prostate can regenerate prostate tissue, Choi et al. (2012) performed lineage tracing on both basal and luminal cells in the normal adult murine prostate (Choi et al. 2012). Using a K14 promoter driving expression of Cre to label basal cells and their progeny and a K8 promoter driving Cre to mark luminal cells and their progeny, Xin and colleagues found that both basal and luminal cells are predominantly self-sustained lineages, presumably due to the coexistence of distinct unipotent lineage-restricted stem cells (Choi et al. 2012). Even after serial castration and regeneration, basal cells only gave rise to new basal cells, while luminal cells only gave rise to new luminal cells. A discussion of seemingly conflicting results (as basal cells and CARN cells are multipotent upon transplantation, but lineage-marked basal and luminal cells are unipotent) is included in Sect. 2.4.

## 2.3.3 *Markers of Isolated Basal Stem Cells in Dissociated Human Prostate Tissues*

Using in vitro assays, it was shown that basal cells from the human prostate can give rise to luminal-like cells in vitro, suggesting a linear relationship between stem cells that reside within the basal layer and their luminal progeny (Robinson et al. 1998).

Collins et al. (2001) demonstrated that CD44+ basal cells from human prostate specimens expressing high levels of alpha-2 integrin preferentially adhere to collagen and form colonies on extracellular matrix-coated plates (Collins et al. 2001). Human prostate basal colony-forming cells can be further enriched in the CD133+ subset when grown on ECM proteins and with irradiated mouse embryonic fibroblasts as feeder cells (Richardson et al. 2004). In vivo, alpha2+ and CD133+ basal cells from the human prostate can generate epithelial structures at a low efficiency (Collins et al. 2001; Richardson et al. 2004). Human prostate basal cells isolated based on mouse prostate stem cell markers Trop2 and CD49f can form spheres at an average rate of almost 1/3, demonstrating significant progenitor activity within the phenotypic fraction Trop2+ CD49f$^{hi}$ (Goldstein et al. 2008). Trop2 and CD49f isolate basal cells that can generate glands upon transplantation into immune-deficient mice that are indistinguishable from primary human prostate tubules (Goldstein et al. 2010). These collective data show that human prostate basal cells show stem cell activity in both in vitro and in vivo assays.

## 2.4 Unresolved Questions for Future Research

As described in Sect. 2.3.2.2, lineage tracing in the adult mouse prostate demonstrated that basal cells give rise to basal cells and luminal cells give rise to luminal cells in a unipotent manner (Choi et al. 2012). Using a dissociated cell tissue-regeneration assay, numerous groups have identified cells capable of multi-lineage differentiation capacity (Burger et al. 2005, 2009; Xin et al. 2005, 2007; Lawson et al. 2007, 2010; Goldstein et al. 2008, 2010; Leong et al. 2008; Wang et al. 2009). Do these assays (tissue recombination using embryonic mesenchyme vs. lineage tracing) measure different activities?

Experiments in the mouse skin indicate that hair follicle bulge stem cells are capable of generating all epidermal lineages upon transplantation, which mimics a wound healing type of response, but they only give rise to hair follicles under normal conditions in the intact skin by lineage tracing (Blanpain and Fuchs 2009). Mouse mammary gland stem cells, identified based on expression of basal cell-surface markers such as high levels of integrins CD49f and CD29, can reconstitute an entire mammary gland upon transplantation into a cleared fat pad (Shackleton et al. 2006; Stingl et al. 2006). However this multi-lineage differentiation capacity is not observed in the intact postpubertal adult mouse mammary gland using lineage tracing tools, as unipotent basal stem cells only give rise to basal cells and lineage-restricted luminal cells are limited to generating adult luminal cells (Van Keymeulen et al. 2011). Interestingly, all adult mouse mammary cells derive from a common embryonic precursor cell, marked by keratin 14 (Van Keymeulen et al. 2011). These collective data suggest that in the prostate and other epithelial tissues, a transplantation approach may push unipotent stem cells toward a more primitive multipotent state, most similar to early development. The use of embryonic mesenchyme in the prostate-regeneration assay may aid adult cells in adopting this embryonic-like fate.

Besides the predominant epithelial cell types, basal and luminal, rare neuroendocrine cells are also found in the developing and adult prostate. However the stem cell that gives rise to neuroendocrine cells remains uncertain. Given the presence of neuroendocrine cells outside of the epithelial glands in the developing human urogenital sinus region, Aumuller et al. (1999) proposed that neuroendocrine cells are derived from the neural crest or ectodermal lineage (Aumuller et al. 1999). However studies by our group and others showed that labeled basal stem cells (Goldstein et al. 2008) or labeled CARN cells (Wang et al. 2009) can give rise to labeled neuroendocrine cells, indicating that they can derive from an endodermal origin in the prostate epithelium.

Through the identification of a panel of cell-surface markers including CD49f, Dick and colleagues have recently demonstrated the purification of single human hematopoietic stem cells capable of long-term engraftment and multi-lineage differentiation (Notta et al. 2011). Isolating stem cells to such a high degree of purity allows for the investigation of their unique properties. Future studies will be necessary to determine whether mouse or human prostate stem cells can be purified to such a degree. Given the emerging role of tissue stem cells in the initiation of mouse (Wang et al. 2006, 2009; Lawson et al. 2010; Choi et al. 2012) and human (Goldstein et al. 2010; Taylor et al. 2012) prostate cancer, identification of unique targets and pathways regulating stem cells will be useful for the future detection and elimination of stem cells in malignant transformation.

## References


Abate-Shen C, Shen MM (2000) Molecular genetics of prostate cancer. Genes Dev 14:2410–2434

Asselin-Labat ML, Vaillant F, Sheridan JM, Pal B, Wu D, Simpson ER, Yasuda H, Smyth GK, Martin TJ, Lindeman GJ et al (2010) Control of mammary stem cell function by steroid hormone signalling. Nature 465:798–802

Aumuller G, Leonhardt M, Janssen M, Konrad L, Bjartell A, Abrahamsson PA (1999) Neurogenic origin of human prostate endocrine cells. Urology 53:1041–1048

Barclay WW, Axanova LS, Chen W, Romero L, Maund SL, Soker S, Lees CJ, Cramer SD (2008) Characterization of adult prostatic progenitor/stem cells exhibiting self-renewal and multilineage differentiation. Stem Cells 26:600–610

Barker N, van Es JH, Kuipers J, Kujala P, van den Born M, Cozijnsen M, Haegebarth A, Korving J, Begthel H, Peters PJ et al (2007) Identification of stem cells in small intestine and colon by marker gene Lgr5. Nature 449:1003–1007

Barrandon Y, Green H (1987) Three clonal types of keratinocyte with different capacities for multiplication. Proc Natl Acad Sci USA 84:2302–2306

Beauchamp JR, Heslop L, Yu DS, Tajbakhsh S, Kelly RG, Wernig A, Buckingham ME, Partridge TA, Zammit PS (2000) Expression of CD34 and Myf5 defines the majority of quiescent adult skeletal muscle satellite cells. J Cell Biol 151:1221–1234

Berardi AC, Wang A, Levine JD, Lopez P, Scadden DT (1995) Functional isolation and characterization of human hematopoietic stem cells. Science (New York, NY) 267:104–108

Blanpain C, Fuchs E (2009) Epidermal homeostasis: a balancing act of stem cells in the skin. Nat Rev Mol Cell Biol 10:207–217

Blum R, Gupta R, Burger PE, Ontiveros CS, Salm SN, Xiong X, Kamb A, Wesche H, Marshall L, Cutler G et al (2009) Molecular signatures of prostate stem cells reveal novel signaling pathways and provide insights into prostate cancer. PLoS One 4:e5722

Bonkhoff H, Wernert N, Dhom G, Remberger K (1991) Basement membranes in fetal, adult normal, hyperplastic and neoplastic human prostate. Virchows Archiv 418:375–381

Burger PE, Xiong X, Coetzee S, Salm SN, Moscatelli D, Goto K, Wilson EL (2005) Sca-1 expression identifies stem cells in the proximal region of prostatic ducts with high capacity to reconstitute prostatic tissue. Proc Natl Acad Sci USA 102:7180–7185

Burger PE, Gupta R, Xiong X, Ontiveros CS, Salm SN, Moscatelli D, Wilson EL (2009) High aldehyde dehydrogenase activity: a novel functional marker of murine prostate stem/progenitor cells. Stem Cells 27:2220–2228

Choi N, Zhang B, Zhang L, Ittmann M, Xin L (2012) Adult murine prostate basal and luminal cells are self-sustained lineages that can both serve as targets for prostate cancer initiation. Cancer Cell 21:253–265

Collins AT, Habib FK, Maitland NJ, Neal DE (2001) Identification and isolation of human prostate epithelial stem cells based on alpha(2)beta(1)-integrin expression. J Cell Sci 114: 3865–3872

Cotsarelis G, Cheng SZ, Dong G, Sun TT, Lavker RM (1989) Existence of slow-cycling limbal epithelial basal cells that can be preferentially stimulated to proliferate: implications on epithelial stem cells. Cell 57:201–209

Cunha GR, Lung B (1978) The possible influence of temporal factors in androgenic responsiveness of urogenital tissue recombinants from wild-type and androgen-insensitive (Tfm) mice. J Exp Zool 205:181–193

Dontu G, Abdallah WM, Foley JM, Jackson KW, Clarke MF, Kawamura MJ, Wicha MS (2003) In vitro propagation and transcriptional profiling of human mammary stem/progenitor cells. Genes Dev 17:1253–1270

Emonard H, Calle A, Grimaud JA, Peyrol S, Castronovo V, Noel A, Lapiere CM, Kleinman HK, Foidart JM (1987a) Interactions between fibroblasts and a reconstituted basement membrane matrix. J Investig Dermatol 89:156–163

Emonard H, Grimaud JA, Nusgens B, Lapiere CM, Foidart JM (1987b) Reconstituted basement-membrane matrix modulates fibroblast activities in vitro. J Cell Physiol 133:95–102

English HF, Santen RJ, Isaacs JT (1987) Response of glandular versus basal rat ventral prostatic epithelial cells to androgen withdrawal and replacement. Prostate 11:229–242

Feldman BJ, Feldman D (2001) The development of androgen-independent prostate cancer. Nat Rev Cancer 1:34–45

Fong CJ, Sherwood ER, Sutkowski DM, Abu-Jawdeh GM, Yokoo H, Bauer KD, Kozlowski JM, Lee C (1991) Reconstituted basement membrane promotes morphological and functional differentiation of primary human prostatic epithelial cells. Prostate 19:221–235

Fong D, Moser P, Krammel C, Gostner JM, Margreiter R, Mitterer M, Gastl G, Spizzo G (2008a) High expression of TROP2 correlates with poor prognosis in pancreatic cancer. Br J Cancer 99:1290–1295

Fong D, Spizzo G, Gostner JM, Gastl G, Moser P, Krammel C, Gerhard S, Rasse M, Laimer K (2008b) TROP2: a novel prognostic marker in squamous cell carcinoma of the oral cavity. Mod Pathol 21:186–191

Fuchs E, Horsley V (2011) Ferreting out stem cells from their niches. Nat Cell Biol 13:513–518

Garraway IP, Sun W, Tran CP, Perner S, Zhang B, Goldstein AS, Hahm SA, Haider M, Head CS, Reiter RE et al (2009) Human prostate sphere-forming cells represent a subset of basal epithelial cells capable of glandular regeneration in vivo. The Prostate

Garraway IP, Sun W, Tran CP, Perner S, Zhang B, Goldstein AS, Hahm SA, Haider M, Head CS, Reiter RE et al (2010) Human prostate sphere-forming cells represent a subset of basal epithelial cells capable of glandular regeneration in vivo. The Prostate 70: 491–501

Goldstein AS, Lawson DA, Cheng D, Sun W, Garraway IP, Witte ON (2008) Trop2 identifies a subpopulation of murine and human prostate basal cells with stem cell characteristics. Proc Natl Acad Sci USA 105:20882–20887

Goldstein AS, Huang J, Guo C, Garraway IP, Witte ON (2010) Identification of a cell of origin for human prostate cancer. Science (New York, NY) 329:568–571

Goldstein AS, Drake JM, Burnes DL, Finley DS, Zhang H, Reiter RE, Huang J, Witte ON (2011) Purification and direct transformation of epithelial progenitor cells from primary human prostate. Nat Protoc 6:656–667

Hudson DL (2004) Epithelial stem cells in human prostate growth and disease. Prostate Cancer Prostatic Dis 7:188–194

Hudson DL, O'Hare M, Watt FM, Masters JR (2000) Proliferative heterogeneity in the human prostate: evidence for epithelial stem cells. Lab Invest 80:1243–1250

Ikuta K, Weissman IL (1992) Evidence that hematopoietic stem cells express mouse c-kit but do not depend on steel factor for their generation. Proc Natl Acad Sci USA 89:1502–1506

Isaacs JT (1985) Control of cell proliferation and cell death in the normal and neoplastic prostate: a stem cell model. In: Rodgers CH et al (eds) Benign prostatic hyperplasia. Department of Health and Human Services, Washington, DC, pp 85–94

Jaks V, Barker N, Kasper M, van Es JH, Snippert HJ, Clevers H, Toftgard R (2008) Lgr5 marks cycling, yet long-lived, hair follicle stem cells. Nat Genet 40:1291–1299

Kim CF, Jackson EL, Woolfenden AE, Lawrence S, Babar I, Vogel S, Crowley D, Bronson RT, Jacks T (2005) Identification of bronchioalveolar stem cells in normal lung and lung cancer. Cell 121:823–835

Kobayashi H, Minami Y, Anami Y, Kondou Y, Iijima T, Kano J, Morishita Y, Tsuta K, Hayashi S, Noguchi M (2010) Expression of the GA733 gene family and its relationship to prognosis in pulmonary adenocarcinoma. Virchows Arch 457:69–76

Lakso M, Sauer B, Mosinger B Jr, Lee EJ, Manning RW, Yu SH, Mulder KL, Westphal H (1992) Targeted oncogene activation by site-specific recombination in transgenic mice. Proc Natl Acad Sci USA 89:6232–6236

Lavker RM, Sun TT (2000) Epidermal stem cells: properties, markers, and location. Proc Natl Acad Sci USA 97:13473–13475

Lawson DA, Witte ON (2007) Stem cells in prostate cancer initiation and progression. J Clin Invest 117:2044–2050

Lawson DA, Xin L, Lukacs RU, Cheng D, Witte ON (2007) Isolation and functional characterization of murine prostate stem cells. Proc Natl Acad Sci USA 104:181–186

Lawson DA, Zong Y, Memarzadeh S, Xin L, Huang J, Witte ON (2010) Basal epithelial stem cells are efficient targets for prostate cancer initiation. Proc Natl Acad Sci USA 107(6): 2610–2615

Leong KG, Wang BE, Johnson L, Gao WQ (2008) Generation of a prostate from a single adult stem cell. Nature 456:804–808

Lim E, Vaillant F, Wu D, Forrest NC, Pal B, Hart AH, Asselin-Labat ML, Gyorki DE, Ward T, Partanen A et al (2009) Aberrant luminal progenitors as the candidate target population for basal tumor development in BRCA1 mutation carriers. Nat Med 15:907–913

Litvinov IV, Vander Griend DJ, Xu Y, Antony L, Dalrymple SL, Isaacs JT (2006) Low-calcium serum-free defined medium selects for growth of normal prostatic epithelial stem cells. Cancer Res 66:8598–8607

Lukacs RU, Lawson DA, Xin L, Zong Y, Garraway I, Goldstein AS, Memarzadeh S, Witte ON (2008) Epithelial stem cells of the prostate and their role in cancer progression. Cold Spring Harb Symp Quant Biol 73:491–502

Lukacs RU, Goldstein AS, Lawson DA, Cheng D, Witte ON (2010a) Isolation, cultivation and characterization of adult murine prostate stem cells. Nat Protoc 5:702–713

Lukacs RU, Memarzadeh S, Wu H, Witte ON (2010b) Bmi-1 is a crucial regulator of prostate stem cell self-renewal and malignant transformation. Cell Stem Cell 7:682–693

Manova K, Nocka K, Besmer P, Bachvarova RF (1990) Gonadal expression of c-kit encoded at the W locus of the mouse. Development 110:1057–1069

Muhlmann G, Spizzo G, Gostner J, Zitt M, Maier H, Moser P, Gastl G, Muller HM, Margreiter R, Ofner D et al (2009) TROP2 expression as prognostic marker for gastric carcinoma. J Clin Pathol 62:152–158

Mulholland DJ, Xin L, Morim A, Lawson D, Witte O, Wu H (2009) Lin-Sca-1+CD49fhigh stem/progenitors are tumor-initiating cells in the Pten-null prostate cancer model. Cancer Res 69:8555–8562

Nguyen DT, Dey A, Lang RJ, Ventura S, Exintaris B (2011) Contractility and pacemaker cells in the prostate gland. J Urol 185:347–351

Notta F, Doulatov S, Laurenti E, Poeppl A, Jurisica I, Dick JE (2011) Isolation of single human hematopoietic stem cells capable of long-term multilineage engraftment. Science (New York, NY) 333:218–221

Ogawa M (1993) Differentiation and proliferation of hematopoietic stem cells. Blood 81: 2844–2853

Ohmachi T, Tanaka F, Mimori K, Inoue H, Yanaga K, Mori M (2006) Clinical significance of TROP2 expression in colorectal cancer. Clin Cancer Res 12:3057–3063

Okada H, Tsubura A, Okamura A, Senzaki H, Naka Y, Komatz Y, Morii S (1992) Keratin profiles in normal/hyperplastic prostates and prostate carcinoma. Virchows Archiv 421:157–161

Ontiveros CS, Salm SN, Wilson EL (2008) Axin2 expression identifies progenitor cells in the murine prostate. Prostate 68:1263–1272

Peng W, Bao Y, Sawicki JA (2011) Epithelial cell-targeted transgene expression enables isolation of cyan fluorescent protein (CFP)-expressing prostate stem/progenitor cells. Transgenic Res 20:1073–1086

Ploemacher RE, Brons RH (1989) Separation of CFU-S from primitive cells responsible for reconstitution of the bone marrow hemopoietic stem cell compartment following irradiation: evidence for a pre-CFU-S cell. Exp Hematol 17:263–266

Reynolds BA, Weiss S (1996) Clonal and population analyses demonstrate that an EGF-responsive mammalian embryonic CNS precursor is a stem cell. Dev Biol 175:1–13

Richardson GD, Robson CN, Lang SH, Neal DE, Maitland NJ, Collins AT (2004) CD133, a novel marker for human prostatic epithelial stem cells. J Cell Sci 117:3539–3545

Robinson EJ, Neal DE, Collins AT (1998) Basal cells are progenitors of luminal cells in primary cultures of differentiating human prostatic epithelium. Prostate 37:149–160

Rock JR, Onaitis MW, Rawlins EL, Lu Y, Clark CP, Xue Y, Randell SH, Hogan BL (2009) Basal cells as stem cells of the mouse trachea and human airway epithelium. Proc Natl Acad Sci USA 106:12771–12775

Shackleton M, Vaillant F, Simpson KJ, Stingl J, Smyth GK, Asselin-Labat ML, Wu L, Lindeman GJ, Visvader JE (2006) Generation of a functional mammary gland from a single stem cell. Nature 439:84–88

Shafik A, Shafik I, el-Sibai O (2005) Identification of c-kit-positive cells in the human prostate: the interstitial cells of Cajal. Arch Androl 51:345–351

Shahi P, Seethammagari MR, Valdez JM, Xin L, Spencer DM (2011) Wnt and Notch pathways have interrelated opposing roles on prostate progenitor cell proliferation and differentiation. Stem Cells (Dayton, Ohio) 29:678–688

Shen MM, Abate-Shen C (2010) Molecular genetics of prostate cancer: new prospects for old challenges. Genes Dev 24:1967–2000

Shen MM, Wang X, Economides KD, Walker D, Abate-Shen C (2008) Progenitor cells for the prostate epithelium: roles in development, regeneration, and cancer. Cold Spring Harb Symp Quant Biol 73:529–538

Shi X, Gipp J, Bushman W (2007) Anchorage-independent culture maintains prostate stem cells. Dev Biol 312:396–406

Slack JM (2000) Stem cells in epithelial tissues. Science (New York, NY) 287:1431–1433

Soriano P (1999) Generalized lacZ expression with the ROSA26 Cre reporter strain. Nat Genet 21:70–71

Spangrude GJ, Heimfeld S, Weissman IL (1988) Purification and characterization of mouse hematopoietic stem cells. Science (New York, NY) 241:58–62

Stingl J, Eirew P, Ricketson I, Shackleton M, Vaillant F, Choi D, Li HI, Eaves CJ (2006) Purification and unique properties of mammary epithelial stem cells. Nature 439:993–997

Taylor RA, Toivanen R, Frydenberg M, Pedersen J, Harewood L, Australian Prostate Cancer B, Collins AT, Maitland NJ, Risbridger GP (2012) Human epithelial basal cells are cells of origin of prostate cancer, independent of CD133 status. Stem Cells (Dayton, Ohio) 30:1087–1096

Thorgeirsson SS (1996) Hepatic stem cells in liver regeneration. FASEB J 10:1249–1256

Tsujimura A, Koikawa Y, Salm S, Takao T, Coetzee S, Moscatelli D, Shapiro E, Lepor H, Sun TT, Wilson EL (2002) Proximal location of mouse prostate epithelial stem cells: a model of prostatic homeostasis. J Cell Biol 157:1257–1265

Uzgare AR, Xu Y, Isaacs JT (2004) In vitro culturing and characteristics of transit amplifying epithelial cells from human prostate tissue. J Cell Biochem 91:196–205

Van der Aa F, Roskams T, Blyweert W, De Ridder D (2003) Interstitial cells in the human prostate: a new therapeutic target? Prostate 56:250–255

Van Keymeulen A, Rocha AS, Ousset M, Beck B, Bouvencourt G, Rock J, Sharma N, Dekoninck S, Blanpain C (2011) Distinct stem cells contribute to mammary gland development and maintenance. Nature 479:189–193

van Leenders G, Dijkman H, Hulsbergen-van de Kaa C, Ruiter D, Schalken J (2000) Demonstration of intermediate cells during human prostate epithelial differentiation in situ and in vitro using triple-staining confocal scanning microscopy. Lab Invest 80:1251–1258

van Leenders GJ, Gage WR, Hicks JL, van Balken B, Aalders TW, Schalken JA, De Marzo AM (2003) Intermediate cells in human prostate epithelium are enriched in proliferative inflammatory atrophy. Am J Pathol 162:1529–1537

Wang S, Garcia AJ, Wu M, Lawson DA, Witte ON, Wu H (2006) Pten deletion leads to the expansion of a prostatic stem/progenitor cell subpopulation and tumor initiation. Proc Natl Acad Sci USA 103:1480–1485

Wang X, Kruithof-de Julio M, Economides KD, Walker D, Yu H, Halili MV, Hu YP, Price SM, Abate-Shen C, Shen MM (2009) A luminal epithelial stem cell that is a cell of origin for prostate cancer. Nature 461:495–500

Welm BE, Tepera SB, Venezia T, Graubert TA, Rosen JM, Goodell MA (2002) Sca-1(pos) cells in the mouse mammary gland represent an enriched progenitor cell population. Dev Biol 245:42–56

Xin L, Ide H, Kim Y, Dubey P, Witte ON (2003) In vivo regeneration of murine prostate from dissociated cell populations of postnatal epithelia and urogenital sinus mesenchyme. Proc Natl Acad Sci USA 100(Suppl 1):11896–11903

Xin L, Lawson DA, Witte ON (2005) The Sca-1 cell surface marker enriches for a prostate-regenerating cell subpopulation that can initiate prostate tumorigenesis. Proc Natl Acad Sci USA 102:6942–6947

Xin L, Lukacs RU, Lawson DA, Cheng D, Witte ON (2007) Self-renewal and multilineage differentiation in vitro from murine prostate stem cells. Stem Cells(Dayton, Ohio) 25:2760–2769

Xue Y, Smedts F, Debruyne FM, de la Rosette JJ, Schalken JA (1998) Identification of intermediate cell types by keratin expression in the developing human prostate. Prostate 34:292–301

Zhang L, Valdez JM, Zhang B, Wei L, Chang J, Xin L (2011) ROCK inhibitor Y-27632 suppresses dissociation-induced apoptosis of murine prostate stem/progenitor cells and increases their cloning efficiency. PLoS One 6:e18271

# Chapter 3
# Prostate Cancer Stem Cells: A Brief Review

**Xin Chen and Dean G. Tang**

**Abstract** Human cancers have been shown to harbor stem cell-like cells called cancer stem cells (CSCs). These cells are thought to be endowed with indefinite self-renewal ability and believed to be involved in tumor initiation, promotion, progression, metastasis, and therapy resistance. Prostate cancers (PCa) have also been shown to contain CSCs. Here we briefly review the literature reports of CSCs in various tumor systems. We then summarize studies of prostate CSCs (PCSCs) in human cancers and mouse models and discuss their respective limitations. We further discuss the current controversies with respect to identifying the cell of origin for PCa. Elucidating the unique characteristics of PCSCs will enhance our understanding of the mechanisms underlying the emergence of castration-resistant disease and may provide new opportunities for developing therapeutics that target the recurrent PCa.

## Abbreviations

| | |
|---|---|
| ADT | Androgen-deprivation therapy |
| ALDH | Aldehyde dehydrogenase |
| AML | Acute myeloid leukemia |
| AR | Androgen receptor |
| CSCs | Cancer stem cells |
| HPCa | Human prostate cancer |
| miRNAs | MicroRNAs |

X. Chen, Ph.D • D.G. Tang, Ph.D. (✉)
Department of Molecular Carcinogenesis, The University of Texas M.D. Anderson Cancer Center, Science Park, Smithville, TX 78957, USA

Program in Molecular Carcinogenesis, Graduate School of Biomedical Sciences (GSBS), Houston, TX 77030, USA
e-mail: dtang@mdanderson.org

S.D. Cramer (ed.), *Stem Cells and Prostate Cancer*, DOI 10.1007/978-1-4614-6498-3_3, 

| | |
|---|---|
| PCa | Prostate cancer |
| PCSCs | Prostate cancer stem cells |
| SCs | Stem cells |
| SP | Side population |
| TICs | Tumor-initiating cells |

## 3.1 Introduction

Prostate cancer (PCa) is one of the most common cancers affecting men in the Western world and the second leading cause for cancer-related death in American males. There are ~241,740 estimated new cases and ~28,170 estimated deaths in the USA in 2012 (Siegel et al. 2012). Radical prostatectomy is commonly utilized for treatment of early-stage PCa, whereas androgen-deprivation therapy (ADT) is the mainstay treatment for advanced PCa. Unfortunately, almost all treated patients eventually fail ADT and develop castration-resistant PCa, at which stage the disease becomes incurable and fatal.

PCa is a multifocal, heterogeneous disease. The exact etiology for PCa development is not clearly understood. Cancer cell heterogeneity in general is explained by two models: (stochastic) clonal evolution model or the cancer stem cell (CSC) model. The classic clonal evolution model has postulated that all cancer cells are tumorigenic, and therapies need to eliminate as many tumor cells as possible to cure the disease. On the other hand, considerable evidence has shown that many cancers may contain a population of stem cell-like cells that is capable of phenotypic diversification and functional maturation, thus generating heterogeneous cancer cell progeny. These CSCs are believed to be responsible for tumor initiation, formation, progression, relapse, metastasis, and therapy resistance. Importantly, these CSCs have the ability to self-renew and generate diverse bulk cells that constitute the tumor (Visvader and Lindeman 2008). In PCa, CSCs are also posited to be cells that mediate PCa recurrence upon androgen deprivation (Feldman and Feldman 2001; Sharifi et al. 2006). It should be noted that the clonal evolution and CSC models may not be mutually exclusive in explaining tumor cell heterogeneity (Tang 2012).

## 3.2 Cancer Stem Cells: Early Findings, Current Studies, Functional Definition, and Purported Characteristics

In 1937, Furth and Kahn (1937) performed quantitative assays in leukemia cell lines and found that a single murine leukemic cell was able to reinitiate a tumor in a mouse, providing early evidence for CSCs. In 1960, Pierce et al. (1960) observed that undifferentiated teratocarcinoma cells had higher mitotic activity and suggested that these cells might represent teratocarcinoma stem cells. This work was followed by Bruce and Van Der Gaag (1963), in 1963, to measure clonogenic potential of tumor cells that are capable of initiating tumor development. Several groups in the 1960s and 1970s revealed functional heterogeneity in hematological tumors (Clarkson et al. 1970; Clarkson 1969; Killmann et al. 1963), and the work (together with others') suggested that a fraction of proliferative leukemic cells can replenish the bulk leukemic blasts and result in leukemia in vivo. However, identification of a true CSC population was not successful until the 1990s, when Dick and colleagues provided solid evidence that most subtypes of acute myeloid leukemia (AML) are organized as a hierarchy and only the $CD34^{+}CD38^{-}$ leukemic stem cells have the ability to serially reconstitute AML in immunodeficient mice (Lapidot et al. 1994; Bonnet and Dick 1997). Clarke and colleagues were the first to report stem cell-like cells in a solid tumor, i.e., $CD44^{+}CD24^{-/lo}$ breast cancer cells (Al-Hajj et al. 2003). They found that as few as 100 of $CD44^{+}CD24^{-/lo}$ breast cancer cells were highly tumorigenic, whereas cells from the other subsets could not regenerate tumors. Since then, putative CSCs have been reported in a variety of human cancers.

CSCs are frequently identified (and characterized) by marker-dependent strategies via flow cytometry sorting for cells positive for specific markers (e.g., CD44) or marker-independent methods, e.g., Aldefluor assay, side population (SP) analysis, and sphere formation assays, combined with limiting-dilution xenotransplantation in immunodeficient mice. For example, in addition to the $CD44^{+}CD24^{-/lo}$ phenotype mentioned above, breast CSCs have also been reported to be enriched in $ALDH^{+}$ (Ginestier et al. 2007), SP (Hirschmann-Jax et al. 2004), or $PHK26^{pos}$ (Pece et al. 2010) cell populations. Brain CSCs have been identified by using the cell-surface marker, i.e., CD133 (Singh et al. 2004) combined with functional analysis such as SP (Bleau et al. 2009) and neurosphere assays (Pastrana et al. 2011). Similarly, colon CSCs have been enriched by using markers such as CD133 (O'Brien et al. 2007; Ricci-Vitiani et al. 2007; Todaro et al. 2007) and CD44 (Dalerba et al. 2007) as well as functional strategies including Aldefluor (Huang et al. 2009) and SP analysis (Inoda et al. 2011). Putative CSCs have now been reported in most other human cancers including those in the lung (Ho et al. 2007; Curtis et al. 2010), pancreas (Li et al. 2007; Lonardo et al. 2011), liver (Yang et al. 2008a; Cairo et al. 2008), head and neck (Prince et al. 2007), stomach (Matsumoto et al. 2009), kidney (Nishizawa et al. 2012; Grange et al. 2011), and ovary (Silva et al. 2011; Meirelles et al. 2011), as well as in melanomas (Schatton et al. 2008; Quintana et al. 2008; Boiko et al. 2010).

How should CSCs be defined? In the strictest sense, a CSC is the *only* cell within the tumor that has the ability to self-renew and generate the heterogeneous lineages of bulk cancer cells (Clarke et al. 2006). None of the reported CSCs would fit this stringent definition. In reality, CSCs are tested in functional assays, in which CSCs are defined as a population of tumor cells that can initiate serially transplantable tumors and can, at least partially, reconstitute the heterogeneity of the original tumors at the histopathological level. The "gold" standard to functionally define CSCs is to show their ability to initiate tumor development in immunodeficient mice and to self-renew by performing serial transplantation assays (Li et al. 2009).

At present, there exist some confusion, controversies, and misunderstandings with respect to characteristics ascribed to CSCs. *First*, the abundance of CSCs may depend on the models studied. It has been mistakenly believed that CSCs represent a small percentage of all tumor cells, which may not always be the case. For example, it was initially estimated that only one in a million human melanoma cells possesses the tumor-initiating ability (Schatton et al. 2008). But it was later reported that frequency of human melanoma CSCs could be as high as one in four melanoma cells when more immunocompromised mice were used (Quintana et al. 2008), suggesting that the frequency of melanoma CSCs varies in different xenograft models. *Second*, normal adult stem cells are characterized by long-term quiescence. Based on this property, CSCs are also assumed to remain at dormancy and retain DNA labels much longer than the non-CSCs, which might help explain why CSCs in some reports are resistant to chemotherapeutics and radiation therapies. Nevertheless, whether or not CSCs are dormant in human tumors has not been clarified and needs further investigation with approved assays, in which a single candidate CSC with a defined quiescence status should be xenotransplanted (Clevers 2011). *Third*, CSCs may or may not originate from normal stem cells. For example, most CSCs are identified via the corresponding normal stem cell marker(s) (Lapidot et al. 1994; Bonnet and Dick 1997), suggesting that CSCs may derive from their normal counterparts. However, CSCs can also derive from restricted progenitors or even differentiated cells. For example, medulloblastoma, the most malignant brain tumor in children, can be initiated in either restricted neuronal progenitors or stem cells (Yang et al. 2008b), suggesting that progenitor/differentiated cells can function as the targets of tumorigenic transformation. *Fourth*, CSCs may or may not always be resistant to therapy. On the one hand, $CD133^+$ glioma stem cells, but not the $CD133^-$ bulk tumor cells, survive after ionizing radiation by preferentially activating the DNA damage checkpoint (Bao et al. 2006). On the other, recent evidence suggests that some CSCs can be targeted by conventional treatments, not all CSCs are therapy resistant, and, vice versa, not all drug-resistant cancer cells are CSCs. For example, we have shown that drug-tolerant (i.e., drug-resistant) residual Du145 PCa cells lacked the expression of the progenitor marker CD44 and actually exhibited reduced tumorigenicity (Yan et al. 2011).

## 3.3 Human Prostate Cancer Stem Cells: Identification, Characterization, Implication, and Challenges

Many studies have reported stem-like PCa cells in long-term cultured PCa cell lines, xenograft models, and primary tumor samples. These reported PCSCs populations seem to be phenotypically divergent, and some have not been rigorously tested in vivo.

Collins et al. (2005) have shown that a small percentage of primary human prostate tumor cells bears the phenotype of $CD44^+\alpha2\beta1^{hi}CD133^+$ and is highly clonogenic and proliferative, although it remains unclear whether the $CD44^+\alpha2\beta1^{hi}CD133^+$ PCa cells can initiate serially transplantable tumors, because such an in vivo experiment was lacking in this study. Using similar flow cytometry-based cell-surface marker strategies, we have reported that in xenograft models (Du145, LAPC4, LAPC9), the $CD44^+$ PCa cells are enriched in tumorigenic and metastatic stem/progenitor cells (Patrawala et al. 2006). We have shown, in subsequent studies, that $CD44^+\alpha2\beta1^+$ PCa cells further enrich CSCs compared to the $CD44^+$ PCa cells (Patrawala et al. 2007). Isaacs's group reported that a small population of $CD133^+$ cells is enriched in several human PCa cell lines (i.e., LNCaP, CWR22Rv1, LAPC4), which can self-renew and produce heterogeneous progeny (Vander Griend et al. 2008). Recently, Rajasekhar et al. (2011) reported a minor subset of human prostate CWR22 xenograft cells expressing TRA-1-60, CD151 and CD166, is capable of reconstituting serially transplantable tumors.

Stem-like PCa cells have also been enriched using functional assays. For example, the LAPC9 SP cells, although rare, are more tumorigenic than non-SP cells (Patrawala et al. 2005). Similarly, aldehyde dehydrogenase (ALDH) $1A1^+$ PCa cells (PC3 and LNCaP), isolated by Aldefluor assay, were shown to be highly clonogenic and tumorigenic, and ALDH1A1 expression can be used to predict patient outcome (Li et al. 2010). Intriguingly, ALDH-positive PCa cells were shown to be enriched in tumor-initiating and metastasis-initiating cells (van den Hoogen et al. 2010). Stem-like PCa cells have also been uncovered using several marker-independent methods. For instance, we have demonstrated that holoclones from cultured PC3 cells are more clonogenic and express higher stem cell-associated molecules than meroclones and paraclones (Li et al. 2008). Importantly, the holoclone cells are able to initiate serially transplantable tumors. Prostaspheres derived from primary human PCa cells harbor cells with self-renewal and clonogenic potential (Guzmán-Ramírez et al. 2009). Recently, we employed a lentiviral reporter system to provide convincing evidence in PCa cell lines, xenografts, and primary PCa that the $PSA^{-/lo}$ PCa cell population harbors CSCs (Qin et al. 2012). The $PSA^{-/lo}$ PCa cells are quiescent and resistant to stresses (e.g., androgen deprivation, chemotherapeutics, and prooxidants). Furthermore, using time-lapse video microscopy, we are the very first to show that a fraction of $PSA^{-/lo}$ cells can undergo asymmetric cell division generating $PSA^+$ cells that constitute the bulk in the tumor. Moreover, $PSA^{-/lo}$ cells can maintain long-term clonogenicity and tumor propagation and mediate castration resistance. As the $PSA^{-/lo}$ cell population is still heterogeneous, we can further enrich tumorigenic and castration-resistant PCa cells using the $ALDH^+CD44^+\alpha2\beta1^{hi}$ phenotype (Qin et al. 2012). This recent study (Qin et al. 2012) provides concrete evidence that human PCa contains CSCs.

What are the mechanisms that regulate human PCSCs? Some evidence suggests that PTEN/PI3K/AKT pathway may play a vital role in maintaining PCSCs, and as a result PI3K signaling is a potential target for PCa treatment (Dubrovska et al. 2009). Combination of PI3K/mTOR inhibitor NVP-BEZ235 and chemotherapeutic drug Taxotere, which target PCSCs and bulk cells, respectively, shows synergistic effects in inhibiting PCa xenograft tumors (Dubrovska et al. 2010). The TRA-1-$60^+$ CSCs showed increased level of NF-κB signaling, and NF-κB inhibitors (i.e., 481407 compound, parthenolide, and celastrol) abrogated sphere formation in a dose-dependent manner (Rajasekhar et al. 2011), suggesting that NF-κB signaling might be involved in PCSC maintenance and functions. By using loss- and gain-of-function studies, our lab has shown that Nanog plays a role in regulating CSC activities and castration resistance and may represent another therapeutic target, especially for CRPC (Jeter et al. 2009, 2011). Finally, we have uncovered that miR-34a is a key negative regulator of PCSCs and PCa metastasis by directly targeting CD44 (Liu et al. 2011). Administration of miR-34a significantly inhibits metastasis and extends the survival of tumor-bearing mice (Liu et al. 2011), indicating that tumor-suppressive microRNAs such as miR-34a could represent potent therapeutic agents for PCa.

Although providing critical information on PCSCs, the above-discussed studies do suffer some limitations. For example, although primary prostate tumors clearly contain highly clonogenic cells (Collins et al. 2005; Guzmán-Ramírez et al. 2009), it remains to be shown whether such cells, freshly purified from patient tumors truly possess enhanced tumorigenic potential and whether primary PCa cells are organized as a hierarchy. These uncertainties are related to the well-known fact that we have yet to find reliable approaches to reconstitute human PCa in immunodeficient mice using freshly purified single tumor cells (Pienta et al. 2008), which is why the majority of the aforementioned studies have been conducted mainly using long-term cultured cell lines and/or long-term xenograft models. In addition, our current CSC assays need improvement and optimization. For instance, many tumor experiments have been performed subcutaneously, which will certainly be different from the authentic PCSC niche in vivo. On the other hand, although orthotopic transplantation may partially ameliorate this problem, these assays will have difficulties in estimating the accurate CSC frequency owing to species incompatibilities. Finally, PCSCs seem to be very heterogeneous, and the interrelationship among the above-mentioned PCSC subpopulations needs to be investigated.

## 3.4 Murine PCSCs: Identification, Characterization, and Implications

Mouse PCa models can overcome the species incompatibility issues that the transplanted human PCa cells may experience in immune compromised hosts. As in the case of human PCSCs, the reported murine PCSCs are also phenotypically divergent. Xin et al. (2005) have demonstrated that the Sca-1$^{+}$ mouse prostate epithelial cells, but not the isogenic Sca-1$^{-}$ cells, are capable of regenerating prostate intraepithelial neoplasia (PIN) lesions after AKT1 lentivirus infection, suggesting that Sca-1$^{+}$/AKT$^{hi}$ cells could become tumor-initiating cells. Mulholland et al. reported a tumor-initiating subpopulation from *Pten*-null PCa model, which bears the phenotype of Lin$^{-}$Sca-1$^{+}$CD49f$^{high}$ (or LSC) and can regenerate tumors in vivo (Mulholland et al. 2009). To date, the LSC subpopulation is one of the few mouse PCSC populations more thoroughly studied.

Molecular mechanisms regulating the maintenance and functions of murine PCSCs remain largely unknown, although some recent exiting evidence starts to emerge. For example, prostaspheres from *Pten/TP53*-null cells possess CSC properties and display increased levels of AKT/mTORC1 and AR pathways, thus implicating such pathways in regulating murine PCSCs (Abou-Kheir et al. 2010). More recently, it has been demonstrated that cooperation of *Pten* loss and *Ras* activation significantly enhances the activity of LSC stem/progenitor cells, leading to their increased EMT (epithelial-mesenchymal transition) and distant metastatic capabilities (Mulholland et al. 2012). In addition to the signaling pathways discussed above, the polycomb group transcriptional repressor Bmi-1 seems to be required for the self-renewal activity in adult murine prostate stem cells (Lukacs et al. 2010). This work may provide another layer of mechanisms, i.e., at the chromatin level, in that Bmi-1 may also regulate PCSCs characteristics such as self-renewal.

Although mouse models have their own advantages for studying CSCs, there also exist apparent limitations. It is unclear to what extent studies in the mouse models can be generalized to human PCa as the mouse and human prostates are quite different. Moreover, considering the heterogeneous nature of CSCs (Tang 2012), it is highly likely that LSCs only represent one population of murine PCSCs, dictated by *Pten* mutation/loss. Future work should determine whether the LSC population also exists in other mouse PCa models, how different genetic mutations may generate distinct CSC types, and the interrelationship among different murine PCSC populations.

## 3.5 Cell of Origin for PCa and CSCs

The potential cell of origin for PCa has attracted much attention and is still under debate. Normal prostatic glands are composed of three distinct types of cells: basal, luminal, and neuroendocrine cells. Basal cells form the basal layer that lines along the basement membrane, and luminal cells constitute the luminal layer that sits above the basal layer and secrete prostatic proteins into the lumen, whereas neuroendocrine cells are dispersed throughout the basal layer and generate neuropeptides and biogenic amines (Abate-Shen and Shen 2000).

It is well known that the untreated clinical prostate tumors contain mostly luminal-like cells expressing AR and cytokeratin 8 with basal-like cells very rare. As such, luminal cells have been presumed to be the cells of origin for PCa. Recent studies (Ma et al. 2005; Korsten et al. 2009) have provided some support to this view. Additionally, using genetic lineage tracing, Wang et al. (2009) have reported that a luminal stem cell population, manifested upon castration and called castration-resistant Nkx3.1-expressing cells (CARNs), can function as the cell of origin for PCa after *Pten* deletion. Using similar strategies, Choi et al. (2012) have shown recently that murine prostatic luminal cells are more susceptible to *Pten* loss-induced PCa, whereas basal cells appear to first need to differentiate into transformation-competent luminal cells in order for *Pten* loss-induced PCa to occur. These two studies (Wang et al. 2009; Choi et al. 2012) were conducted in mouse models, and it is presently unclear whether human prostatic luminal cells can also function as the cells of origin for human PCa. The answer to this latter question awaits the development of culture medium that can propagate $AR^+$ and $PSA^+$

differentiated luminal cells. Interestingly, stem-like cells in BM18 human PCa xenograft express both stem cell markers (i.e., ALDH1A1, Nanog) as well as luminal markers (e.g., NKX3.1 and CK18), but not basal markers, and these cells are selected by castration and are able to regenerate tumors after androgen replacement (Germann et al. 2012), suggesting that stem-like cells with luminal progenitor phenotype in human PCa might represent the cell of origin for CRPC.

In contrast to the above lineage tracing studies (Wang et al. 2009; Choi et al. 2012), tissue recombination/regeneration assays using FACS-purified cells have shown that prostatic basal cells are targets for malignant transformation (Wang et al. 2006; Lawson et al. 2010). When combinations of genetic alterations such as ERG1, constitutively active AKT, and AR are introduced into either purified basal or luminal cells from the mouse prostate, only basal cells seem to be competent for tumorigenic transformation (Lawson et al. 2010). When the same combination of ERG, AKT, and AR is introduced into the isolated human prostatic luminal ($CD49f^{lo}Trop2^{hi}$) or basal ($CD49f^{hi}Trop2^{hi}$) cells, tissue recombination assays in NOD-SCID-IL2R$\gamma^{null}$ (NSG) mice reveal that only basal cells can be transformed, leading to adenocarcinoma that, remarkably, resembles the clinical PCa histology (Goldstein et al. 2010). A recent study also shows that the $\alpha2\beta1integrin^{hi}$, basal-like cells from nontumorigenic BPH-1 cells, when combined with either human cancer-associated fibroblasts or embryonic stroma, can lead to tumor grafts using tissue recombination assays (Taylor et al. 2012).

There are several potential explanations for the seemingly contradictory results from the lineage tracing versus transplantation-based studies. The obvious one is that both luminal and basal epithelial cells can function as the cells of origin for PCa, depending on specific genetic hits and context. The second is technical. Although lineage tracing studies are performed in intact animals, the Cre-mediated tagging efficiency frequently is low and varies with the promoter used, which can confound interpretation of data. On the other hand, there are also several caveats associated with the transplantation-based studies. Transplantation per se may elicit a wound healing response, which could lead to heightened activity of the transplanted cells and skewed results. More important, as of yet we still fail to maintain and propagate fully differentiated luminal prostatic epithelial cells expressing AR and PSA (in human), as all currently used culture media favor only basal cells. Therefore, it remains officially possible that differentiated human prostate luminal cells also can become the targets of tumorigenic transformation. Finally, the different results obtained with the lineage tracing and transplantation studies might also be related to differences between human and mouse cells. It is well established that terminally differentiated human epithelial cells such as PSA-expressing luminal cells generally lack telomerase expression, and in the human prostate most proliferating cells lie in the basal layer (Bonkhoff et al. 1994). In contrast, "differentiated" murine cells have long telomeres and retain high telomerase activity, and in the mouse prostate most proliferative cells reside in the luminal layer (Choi et al. 2012). These species-specific differences could help account for the different results obtained with respect to the cell of origin for PCa.

It should be noted that whether it is lineage tracing or transplantation studies, just because one population of cells can be transformed by specific genetic hit(s) does not necessarily mean that they are the actual cells of origin for human PCa. Also, the potential cells of origin are not to be confused with CSCs, which refer to stem-like cancer cells in the established prostate tumors.

## 3.6 Conclusions and Perspectives

Using different approaches, many studies have demonstrated PCSCs in both mouse models and cultured human PCa cells, xenografts as well as primary patient samples. For example, recent study has shown convincingly that the $PSA^{-/lo}$ PCa cell population harbors stem-like cells that preferentially express stem cell-associated genes, can undergo authentic asymmetric cell division, possess long-term tumor-propagating activity, and can mediate castration resistance (Qin et al. 2012). In mouse PCa models, the best characterized CSC population is the LSCs in *Pten*-less prostate tumors (Mulholland et al. 2009). Both human and mouse PCSCs seem to be phenotypically divergent and future efforts should be directed towards characterizing the interrelationship among various reported CSC populations. Another important breakthrough is urgently needed to develop the assays that allow reliable tumor reconstitutions in mice from purified single primary PCa cells.

Moreover, some recent evidence has implicated certain signaling molecules and pathways such as PTEN/PI3K/AKT, NF-κB, Nanog, and miRNAs in regulating PCSC properties. Better understanding of these and other potential mechanisms is likely to help us develop novel therapeutics to target PCSCs, and eventually benefit PCa patients by preventing recurrence.

Finally, lineage tracing and tissue recombination assays seem to suggest that both luminal and basal cells can function as the targets of tumorigenic transformation. Interpretation of these results should take into account of the experimental system used. Future studies with improved techniques are needed to further elucidate the cells of origin for human PCa, the success of which will be instrumental to stratifying PCa patients and developing personalized and targeted therapeutics.

**Acknowledgements** The work in the authors' lab was supported, in part, by grants from NIH (R21-CA150009 and R01-CA155693-01A1), Department of Defense (W81XWH-11-1-0331), CPRIT (RP120380), and the MDACC Center for Cancer Epigenetics (D.G.T) and by two Center Grants (CCSG-5 P30 CA166672). X. Chen was supported by the Cockrell Fund from the Department of Molecular Carcinogenesis.

## References

Abate-Shen C, Shen MM (2000) Molecular genetics of prostate cancer. Genes Dev 14(19): 2410–2434

Abou-Kheir WG, Hynes PG, Martin PL et al (2010) Characterizing the contribution of stem/progenitor cells to tumorigenesis in the Pten-/-TP53-/- prostate cancer model. Stem Cells 28(12):2129–2140

Al-Hajj M, Wicha MS, Benito-Hernandez A et al (2003) Prospective identification of tumorigenic breast cancer cells. Proc Natl Acad Sci USA 100(7):3983–3988
Bao S, Wu Q, McLendon RE et al (2006) Glioma stem cells promote radioresistance by preferential activation of the DNA damage response. Nature 444(7120):756–760
Bleau AM, Hambardzumyan D, Ozawa T et al (2009) PTEN/PI3K/Akt pathway regulates the side population phenotype and ABCG2 activity in glioma tumor stem-like cells. Cell Stem Cell 4(3):226–235
Boiko AD, Razorenova OV, van de Rijn M et al (2010) Human melanoma-initiating cells express neural crest nerve growth factor receptor CD271. Nature 466(7302):133–137
Bonkhoff H, Stein U, Remberger K (1994) The proliferative function of basal cells in the normal and hyperplastic human prostate. Prostate 24(3):114–118
Bonnet D, Dick JE (1997) Human acute myeloid leukemia is organized as a hierarchy that originates from a primitive hematopoietic cell. Nat Med 3(7):730–737
Bruce WR, Van Der Gaag H (1963) A quantitative assay for the number of murine lymphoma cells capable of proliferation in vivo. Nature 199:79–80
Cairo S, Armengol C, De Reyniès A et al (2008) Hepatic stem-like phenotype and interplay of Wnt/beta-catenin and Myc signaling in aggressive childhood liver cancer. Cancer Cell 14(6):471–484
Choi N, Zhang B, Zhang L et al (2012) Adult murine prostate basal and luminal cells are self-sustained lineages that can both serve as targets for prostate cancer initiation. Cancer Cell 21(2):253–265
Clarke MF, Dick JE, Dirks PB et al (2006) Cancer stem cells—perspectives on current status and future directions: AACR Workshop on cancer stem cells. Cancer Res 66(19):9339–9344
Clarkson BD (1969) Review of recent studies of cellular proliferation in acute leukemia. Natl Cancer Inst Monogr 30:81–120
Clarkson B, Fried J, Strife A et al (1970) Studies of cellular proliferation in human leukemia. 3. Behavior of leukemic cells in three adults with acute leukemia given continuous infusions of 3H-thymidine for 8 or 10 days. Cancer 25(6):1237–1260
Clevers H (2011) The cancer stem cell: premises, promises and challenges. Nat Med. 17 (3):313–339, Review
Collins AT, Berry PA, Hyde C et al (2005) Prospective identification of tumorigenic prostate cancer stem cells. Cancer Res 65(23):10946–10951
Curtis SJ, Sinkevicius KW, Li D et al (2010) Primary tumor genotype is an important determinant in identification of lung cancer propagating cells. Cell Stem Cell 7(1):127–133
Dalerba P, Dylla SJ, Park IK et al (2007) Phenotypic characterization of human colorectal cancer stem cells. Proc Natl Acad Sci USA 104(24):10158–10163
Dubrovska A, Kim S, Salamone RJ et al (2009) The role of PTEN/Akt/PI3K signaling in the maintenance and viability of prostate cancer stem-like cell populations. Proc Natl Acad Sci USA 106(1):268–273
Dubrovska A, Elliott J, Salamone RJ et al (2010) Combination therapy targeting both tumor-initiating and differentiated cell populations in prostate carcinoma. Clin Cancer Res 16(23):5692–5702
Feldman BJ, Feldman D (2001) The development of androgen-independent prostate cancer. Nat Rev Cancer 1(1):34–45
Furth J, Kahn M (1937) The transmission of leukemia of mice with a single cell. Am J Cancer 31:276–282
Germann M, Wetterwald A, Guzmán-Ramirez N et al (2012) Stem-like cells with luminal progenitor phenotype survive castration in human prostate cancer. Stem Cells 30(6):1076–1086
Ginestier C, Hur MH, Charafe-Jauffret E et al (2007) ALDH1 is a marker of normal and malignant human mammary stem cells and a predictor of poor clinical outcome. Cell Stem Cell 1(5):555–567
Goldstein AS, Huang J, Guo C et al (2010) Identification of a cell of origin for human prostate cancer. Science 329(5991):568–571

Grange C, Tapparo M, Collino F et al (2011) Microvesicles released from human renal cancer stem cells stimulate angiogenesis and formation of lung premetastatic niche. Cancer Res 71(15):5346–5356
Guzmán-Ramírez N, Völler M, Wetterwald A et al (2009) In vitro propagation and characterization of neoplastic stem/progenitor-like cells from human prostate cancer tissue. Prostate 69(15):1683–1693
Hirschmann-Jax C, Foster AE, Wulf GG et al (2004) A distinct "side population" of cells with high drug efflux capacity in human tumor cells. Proc Natl Acad Sci USA 101(39):14228–14233
Ho MM, Ng AV, Lam S et al (2007) Side population in human lung cancer cell lines and tumors is enriched with stem-like cancer cells. Cancer Res 67(10):4827–4833
Huang EH, Hynes MJ, Zhang T et al (2009) Aldehyde dehydrogenase 1 is a marker for normal and malignant human colonic stem cells (SC) and tracks SC overpopulation during colon tumorigenesis. Cancer Res 69(8):3382–3389
Inoda S, Hirohashi Y, Torigoe T et al (2011) Cytotoxic T lymphocytes efficiently recognize human colon cancer stem-like cells. Am J Pathol 178(4):1805–1813
Jeter CR, Badeaux M, Choy G et al (2009) Functional evidence that the self-renewal gene NANOG regulates human tumor development. Stem Cells 27(5):993–1005
Jeter CR, Liu B, Liu X et al (2011) NANOG promotes cancer stem cell characteristics and prostate cancer resistance to androgen deprivation. Oncogene 30(36):3833–3845
Killmann SA, Cronkite EP, Robertson JS et al (1963) Estimation of phases of the life cycle of leukemic cells from labeling in human beings in vivo with tritiated thymidine. Lab Invest 12:671–684
Korsten H, Ziel-van der Made A, Ma X et al (2009) Accumulating progenitor cells in the luminal epithelial cell layer are candidate tumor initiating cells in a Pten knockout mouse prostate cancer model. PLoS One 4(5):e5662
Lapidot T, Sirard C, Vormoor J et al (1994) A cell initiating human acute myeloid leukaemia after transplantation into SCID mice. Nature 367(6464):645–648
Lawson DA, Zong Y, Memarzadeh S et al (2010) Basal epithelial stem cells are efficient targets for prostate cancer initiation. Proc Natl Acad Sci USA 107(6):2610–2615
Li C, Heidt DG, Dalerba P et al (2007) Identification of pancreatic cancer stem cells. Cancer Res 67(3):1030–1037
Li H, Chen X, Calhoun-Davis T et al (2008) PC3 human prostate carcinoma cell holoclones contain self-renewing tumor-initiating cells. Cancer Res 68(6):1820–1825
Li H, Jiang M, Honorio S et al (2009) Methodologies in assaying prostate cancer stem cells. Methods Mol Biol 568:85–138
Li T, Su Y, Mei Y et al (2010) ALDH1A1 is a marker for malignant prostate stem cells and predictor of prostate cancer patients' outcome. Lab Invest 90(2):234–244
Liu C, Kelnar K, Liu B et al (2011) The microRNA miR-34a inhibits prostate cancer stem cells and metastasis by directly repressing CD44. Nat Med 17(2):211–215
Lonardo E, Hermann PC, Mueller MT et al (2011) Nodal/Activin signaling drives self-renewal and tumorigenicity of pancreatic cancer stem cells and provides a target for combined drug therapy. Cell Stem Cell 9(5):433–446
Lukacs RU, Memarzadeh S, Wu H et al (2010) Bmi-1 is a crucial regulator of prostate stem cell self-renewal and malignant transformation. Cell Stem Cell 7(6):682–693
Ma X, Ziel-van der Made AC, Autar B et al (2005) Targeted biallelic inactivation of Pten in the mouse prostate leads to prostate cancer accompanied by increased epithelial cell proliferation but not by reduced apoptosis. Cancer Res 65(13):5730–5739
Matsumoto K, Arao T, Tanaka K et al (2009) mTOR signal and hypoxia-inducible factor-1 alpha regulate CD133 expression in cancer cells. Cancer Res 69(18):7160–7164
Meirelles K, Benedict LA, Dombkowski D et al (2011) Human ovarian cancer stem/progenitor cells are stimulated by doxorubicin but inhibited by Mullerian inhibiting substance. Proc Natl Acad Sci USA 109(7):2358–2363
Mulholland DJ, Xin L, Morim A et al (2009) Lin-Sca-1+CD49fhigh stem/progenitors are tumor-initiating cells in the Pten-null prostate cancer model. Cancer Res 69(22):8555–8562

Mulholland DJ, Kobayashi N, Ruscetti M et al (2012) Pten loss and RAS/MAPK activation cooperate to promote EMT and metastasis initiated from prostate cancer stem/progenitor cells. Cancer Res 72(7):1878–1889

Nishizawa S, Hirohashi Y, Torigoe T et al (2012) HSP DNAJB8 controls tumor-initiating ability in renal cancer stem-like cells. Cancer Res 72(11):2844–2854

O'Brien CA, Pollett A, Gallinger S et al (2007) A human colon cancer cell capable of initiating tumour growth in immunodeficient mice. Nature 445(7123):106–110

Pastrana E, Silva-Vargas V, Doetsch F (2011) Eyes wide open: a critical review of sphere-formation as an assay for stem cells. Cell Stem Cell 8(5):486–498

Patrawala L, Calhoun T, Schneider-Broussard R et al (2005) Side population is enriched in tumorigenic, stem-like cancer cells, whereas ABCG2+ and ABCG2- cancer cells are similarly tumorigenic. Cancer Res 65(14):6207–6219

Patrawala L, Calhoun T, Schneider-Broussard R et al (2006) Highly purified CD44+ prostate cancer cells from xenograft human tumors are enriched in tumorigenic and metastatic progenitor cells. Oncogene 25(12):1696–1708

Patrawala L, Calhoun-Davis T, Schneider-Broussard R et al (2007) Hierarchical organization of prostate cancer cells in xenograft tumors: the CD44+alpha2beta1+ cell population is enriched in tumor-initiating cells. Cancer Res 67(14):6796–6805

Pece S, Tosoni D, Confalonieri S et al (2010) Biological and molecular heterogeneity of breast cancers correlates with their cancer stem cell content. Cell 140(1):62–73

Pienta KJ, Abate-Shen C, Agus DB et al (2008) The current state of preclinical prostate cancer animal models. Prostate 68(6):629–639

Pierce GB Jr, Dixon FJ Jr, Verney EL (1960) Teratocarcinogenic and tissue-forming potentials of the cell types comprising neoplastic embryoid bodies. Lab Invest 9:583–602

Prince ME, Sivanandan R, Kaczorowski A et al (2007) Identification of a subpopulation of cells with cancer stem cell properties in head and neck squamous cell carcinoma. Proc Natl Acad Sci USA 104(3):973–978

Qin J, Liu X, Laffin B et al (2012) The PSA(-/lo) Prostate Cancer Cell Population Harbors Self-Renewing Long-Term Tumor-Propagating Cells that Resist Castration. Cell Stem Cell 10(5):556–569

Quintana E, Shackleton M, Sabel MS et al (2008) Efficient tumour formation by single human melanoma cells. Nature 456(7222):593–598

Rajasekhar VK, Studer L, Gerald W et al (2011) Tumour-initiating stem-like cells in human prostate cancer exhibit increased NF-κB signalling. Nat Commun 2:162

Ricci-Vitiani L, Lombardi DG, Pilozzi E et al (2007) Identification and expansion of human colon-cancer-initiating cells. Nature 445(7123):111–115

Schatton T, Murphy GF, Frank NY et al (2008) Identification of cells initiating human melanomas. Nature 451(7176):345–349

Sharifi N, Kawasaki BT, Hurt EM et al (2006) Stem cells in prostate cancer: resolving the castrate-resistant conundrum and implications for hormonal therapy. Cancer Biol Ther 5(8):901–906

Siegel R, Naishadham D, Jemal A (2012) Cancer statistics, 2012. CA Cancer J Clin 62(1):10–29

Silva IA, Bai S, McLean K et al (2011) Aldehyde dehydrogenase in combination with CD133 defines angiogenic ovarian cancer stem cells that portend poor patient survival. Cancer Res 71(11):3991–4001

Singh SK, Hawkins C, Clarke ID et al (2004) Identification of human brain tumour initiating cells. Nature 432(7015):396–401

Tang DG (2012) Understanding cancer stem cell heterogeneity and plasticity. Cell Res 22(3):457–472

Taylor RA, Toivanen R, Frydenberg M et al (2012) Human epithelial basal cells are cells of origin of prostate cancer, independent of CD133 status. Stem Cells 30(6):1087–1096

Todaro M, Alea MP, Di Stefano AB et al (2007) Colon cancer stem cells dictate tumor growth and resist cell death by production of interleukin-4. Cell Stem Cell 1(4):389–402

van den Hoogen C, van der Horst G, Cheung H et al (2010) High aldehyde dehydrogenase activity identifies tumor-initiating and metastasis-initiating cells in human prostate cancer. Cancer Res 70(12):5163–5173

Vander Griend DJ, Karthaus WL, Dalrymple S et al (2008) The role of CD133 in normal human prostate stem cells and malignant cancer-initiating cells. Cancer Res 68(23):9703–9711

Visvader JE, Lindeman GJ (2008) Cancer stem cells in solid tumours: accumulating evidence and unresolved questions. Nat Rev Cancer 8(10):755–768

Wang S, Garcia AJ, Wu M et al (2006) Pten deletion leads to the expansion of a prostatic stem/progenitor cell subpopulation and tumor initiation. Proc Natl Acad Sci USA 103(5):1480–1485

Wang X, Kruithof-de Julio M, Economides KD et al (2009) A luminal epithelial stem cell that is a cell of origin for prostate cancer. Nature 461(7263):495–500

Xin L, Lawson DA, Witte ON (2005) The Sca-1 cell surface marker enriches for a prostate-regenerating cell subpopulation that can initiate prostate tumorigenesis. Proc Natl Acad Sci USA 102(19):6942–6947

Yan H, Chen X, Zhang Q et al (2011) Drug-tolerant cancer cells show reduced tumor-initiating capacity: depletion of CD44 cells and evidence for epigenetic mechanisms. PLoS One 6(9):e24397

Yang ZF, Ho DW, Ng MN et al (2008a) Significance of CD90+ cancer stem cells in human liver cancer. Cancer Cell 13(2):153–166

Yang ZJ, Ellis T, Markant SL et al (2008b) Medulloblastoma can be initiated by deletion of Patched in lineage-restricted progenitors or stem cells. Cancer Cell 14(2):135–145

# Chapter 4
# Cancer Stem Cells Provide New Insights into the Therapeutic Responses of Human Prostate Cancer

**Fiona M. Frame and Norman J. Maitland**

**Abstract** Whilst there has been a dramatic improvement in the survival of men with prostate cancer in the last few decades, we find ourselves at a crossroads, where another significant therapeutic advance is required. Current treatments for prostate cancer including hormone therapy, radiotherapy and chemotherapy all have their place, but result in almost inevitable treatment failure. In this chapter, we describe the role of cancer stem cells in tumour relapse as well as the potential they provide to develop novel treatment strategies. We examine the clinical implications of cancer stem cells as a therapy-resistant pool within prostate tumours and propose three strategies to target both cancer stem cells and bulk tumour cells, namely, combination therapy, differentiation therapy and targeted therapy. This chapter summarises the challenge of designing future therapies taking into account both the heterogeneity of prostate cancers and the resistant cancer stem cells at their core.

## 4.1 Introduction

The existence of the presence of a primitive underlying epithelial tissue stem cell (SC) in prostate has provided the impetus for a new generation of cell and molecular studies on the origins and the very nature of human prostate cancers. Whilst current thinking about the cellular origins, composition and phenotype of prostate tumours is treated elsewhere in this volume, the existence of primitive epithelial cell populations within tumours contributes to intratumoral heterogeneity and its therapeutic consequences, i.e. the development of treatment-resistant tumour clones. Indeed,

F.M. Frame, Ph.D • N.J. Maitland, Ph.D. (✉)
YCR Cancer Research Unit, Department of Biology, University of York, Heslington, North Yorkshire, YO10 5DD, UK
e-mail: n.j.maitland@york.ac.uk

S.D. Cramer (ed.), *Stem Cells and Prostate Cancer*, DOI 10.1007/978-1-4614-6498-3_4, 

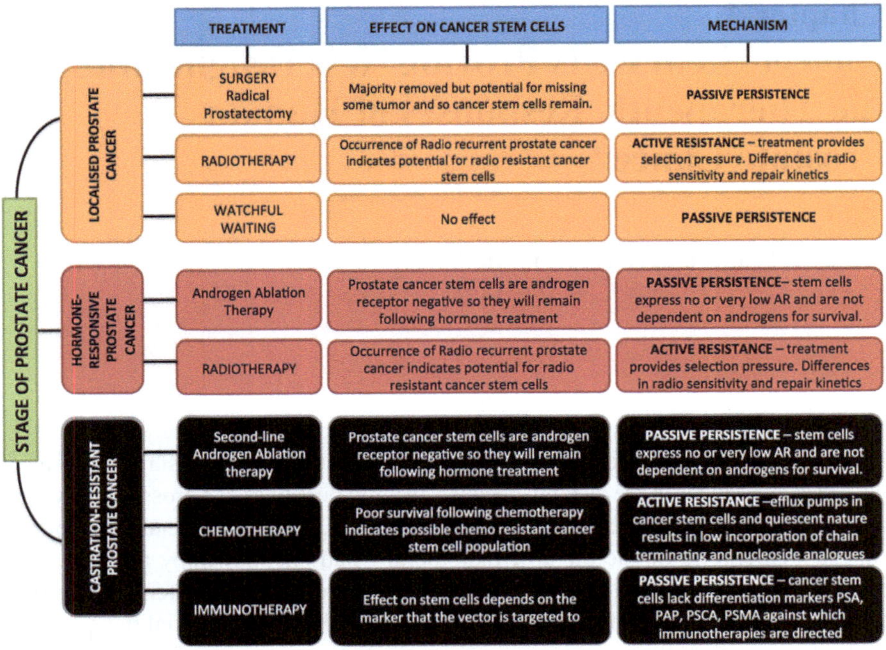


**Fig. 4.1** Therapy resistance in the context of cancer stem cells. The different stages of prostate cancer are represented, ranging in their severity from localised to hormone responsive to castration resistant. Each stage has a range of treatments available and the effect of the treatment on cancer stem cells is listed alongside the mechanism of cancer stem cell maintenance including passive persistence and active resistance

human prostate cancer has justified its reputation as a tumour that is difficult to treat by conventional therapeutic strategies (as summarised in Fig. 4.1).

In the context of organ confined disease, external beam radiotherapy and brachytherapy techniques offer a high degree of disease control, even cure, but with an acknowledged relapse rate of 30% (Ishkanian et al. 2010; Jones 2011; Xiao et al. 2012). In many cases, this is attributable to the pre-existence of metastatic clones, outside the irradiated zone, but recurrence within the prostate is also found. Responses to manipulation of hormone levels, either directly by blocking the androgen receptor or indirectly by (1) stemming the supply of adrenal androgens, (2) preventing activation of testosterone to the more bioactive dihydrotestosterone or (3) the recent introduction of inhibitors of the salvage synthetic pathway for testosterone biosynthesis, are all effective but are clearly time limited in their efficacy (Rehman and Rosenberg 2012; Schroder et al. 2012). However, it is particularly after the failure of such hormone manipulation strategies that the chemoresistance of prostate tumours is most manifest, where survival times for castration-resistant prostate cancer (CRPC) patients rarely exceed 2 years (Kirby et al. 2011).

To justify such rapid failure of potent anti-proliferative treatments, several explanations have been proposed. The emergence of new tumour clones, either spontaneously due to more rapid growth or in a new environment after therapy, can be viewed as a form of adaption and selection. Induction of mutations can be far from random, but the overwhelming pressure comes from the selection process. It is clear that there is considerable clonal evolution of prostate cancers (Ruiz et al. 2011; Kallioniemi and Visakorpi 1996; Cheng et al. 1999). In addition, next-generation DNA/RNA sequencing has indicated that the degree of overall genetic changes in a prostate cancer is somewhat less than in other solid tumours. However, prostate cancer appears to differ from several other common cancer types (e.g. colon), in that the incidences of DNA repair defects and genetic instabilities are frequently lower than expected. The ability to resist the apoptotic consequences of genetic instability does appear to be greater in CRPC, where inactivating mutations in p53 have most recently been confirmed in around 21.4% of cases (Eastham et al. 1995; Navone et al. 1999; Mirchandani et al. 1995). Whilst prostate cancer cells express classical markers of apoptosis resistance, they nevertheless turn over rather rapidly, dividing more frequently than epithelial cells in the normal prostate, and certainly much more frequently than normal luminal epithelial cells, which the bulk of a tumour most closely resembles phenotypically (Limas and Frizelle 1994; Hudson et al. 2001). There are also the common arguments about life span of tumour cells versus the time required to generate the desired mutation, which imply that the tumour originates in relatively long-lived stem/progenitor cells, in contrast to luminal cells, which turn over more rapidly (De Marzo et al. 1998).

Therapy resistance resulting from adaptive mutagenesis in a replicating luminal cell population of prostate tumours is not only appealing but also has some circumstantial evidence in its favour (Germann et al. 2012; Kumar et al. 2011) as shown in Fig. 4.2a. The latter experiments in cell lines and established xenografts, which at least in vitro, seem spectacularly vulnerable to the effects of chemotherapy, did pose the question of just how easy it is to select for the appropriate changes. Recent studies to generate a TMPRSS2-ERG fusion designed under ideal conditions of cell replication, hormone treatment and selection in culture took many months with many millions of starting cells (Bastus et al. 2010; Haffner et al. 2010; Mani et al. 2009). This is at best equivalent to several years' selection in patients. However, chemotherapy-resistant cell clones arise in vivo within several months of treatment instigation (Marin-Aguilera et al. 2012; Mezynski et al. 2012; Seruga et al. 2011) in the clinic. The other consideration must be the phenotype of the most aggressive CRPCs, which is frequently neuroendocrine (Matei et al. 2012; Marcu et al. 2010). How simple is a 'transdifferentiation' from a replicating luminal cancer cell into such a NE phenotype? Clearly, it can be achieved in established cell lines such as LNCaP (Zelivianski et al. 2001), but are such changes so likely in vivo? Direct DNA sequencing has now indicated that patient tumours have many fewer genetic changes than established cell lines and retain some heterogeneity. The most common mechanism of castration resistance is not, as previously proposed, specificity-changing mutations in the androgen receptor sequence (Hay and McEwan 2012) (made easier by the single copy of AR on the X chromosome in human male cells) but rather an

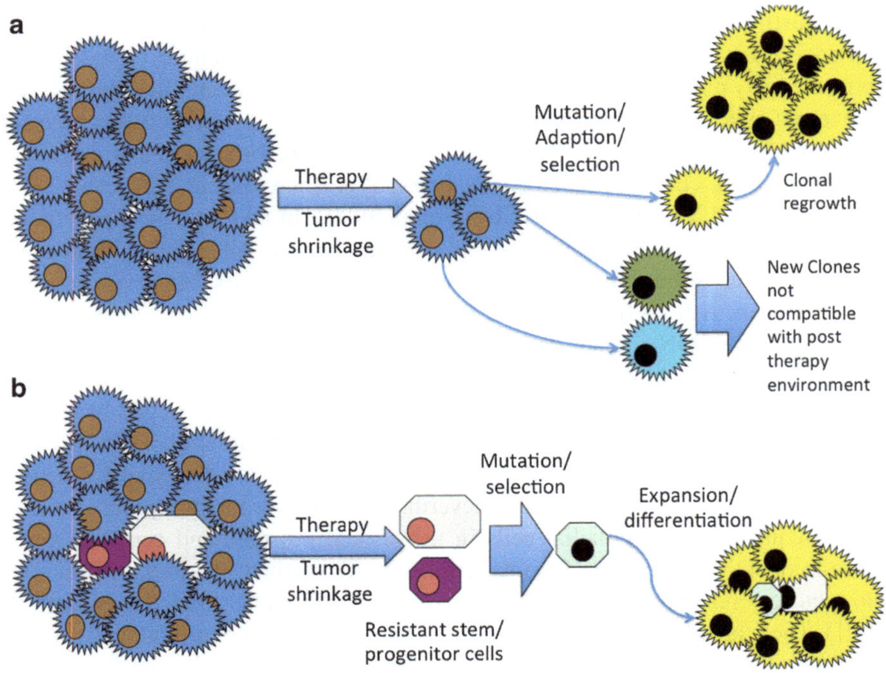


**Fig. 4.2** Adaptive mutagenesis versus stem cell mutation and selection in therapy resistance. (**a**) Adaptive mutagenesis: changes are induced in many tumour cells in response to therapy leading to emergence of a successful relapse originating in a single clone. (**b**) Stem cell mutation and expansion arises from stem/progenitor cells which can expand and differentiate to reconstitute the relapsed tumour mass

amplification of AR gene, resulting in an overexpression of the AR protein under castrate conditions (Koivisto et al. 1997). This has also been considered surprising, but perhaps should not be so, since on consideration of another common chemotherapeutic resistance mechanism, against methotrexate, it was shown as long ago as the 1980s (Stark and Wahl 1984) that resistant tumours had also amplified the substrate binding/metabolising (dihydrofolate reductase) gene (Sharifi et al. 2006).

An alternative mechanism for the development of treatment-resistant clones can be derived from the existence of an underlying stem/progenitor population within the heterogeneous tumour mass (Fig. 4.2b). This is also not a novel or recent idea, both in leukaemia (Lapidot et al. 1994), solid tumours (Visvader and Lindeman 2008) and indeed in prostate cancer (Collins et al. 2005), where Isaacs and Coffey proposed just such an hypothesis almost 30 years ago (Isaacs and Coffey 1981, 1989; Kyprianou and Isaacs 1988). However, these historical hypotheses were generated in scientific times when the ability to purify discrete cell populations was limited by the available technology. With an increasing use of cell sorting techniques, more homogeneous populations resulting from fractionation of cell cultures and tissues resulted in a rediscovery of the cancer stem cell hypothesis. Better knowledge of the cancer stem cell phenotype suggested the molecular mechanisms whereby cancer stem cells could indeed provide the therapy-resistant fraction, from which a new recurrent clone could arise. In 2005, we proposed that a rare cell with cancer stem cell-like properties could provide the reservoir for therapy resistance in prostate cancer (Maitland and Collins 2005). A number of subsequent publications have endorsed this idea, mainly with studies of stem-like cells from prostate cell lines, but also from our own studies on fractionated cells and primary cultures from human subjects (Frame et al. 2010; Goldstein et al. 2010a, b; Maitland et al. 2011; Patrawala et al. 2006, 2007; Trerotola et al. 2010; Lawson et al. 2010).

## 4.2 Success and Failure with Current Prostate Cancer Treatment

There has been great progress in recent decades in terms of detection and diagnosis of prostate cancer. The 5-year survival rate has risen in the last 20 years, taking into account all stages, with mortality declining by about 40% (Siegel et al. 2012; Etzioni et al. 2012). This relates to early diagnosis resulting from PSA testing, refined surgical techniques, successful radiotherapy, improved awareness and evolving imaging techniques. Men are cured of prostate cancer by radical prostatectomy and/or successful radiotherapy, either external beam or brachytherapy. In addition, there is the acknowledgement that the PSA test is less than perfect and judicious use of the watchful waiting/active surveillance approach has prevented unnecessary surgery for a subset of patients, typically older patients with low-grade cancers (Drachenberg 2000). However, there is still an unacceptably high rate of failure of radiotherapy, which consists of either nonresponsive tumours or recurrent tumours. In addition, although androgen ablation therapy is initially very effective and is typically initiated once the tumour has escaped the prostate capsule, hormone resistance almost inevitably emerges. This is followed by the more rapid failure of chemotherapy used to treat castration-resistant tumours. Indeed, the real therapeutic challenge at the moment is with patients harbouring metastatic castration-resistant disease, for whom untreated survival is typically 9–13 months (Kirby et al. 2011).

## 4.3 Therapy Resistance in the Context of Cancer Stem Cells

Illustrated in Fig. 4.1 are the stages of prostate cancer, together with the typical current treatments used against each stage. Considering the prostate cancer(s) as a heterogeneous tumour with a core of cancer stem cells, the figure also describes the potential effect of each of these treatments on the cancer stem cell population. Essentially, the effects are divided into *active resistance* and *passive persistence*. In all cases, the stem cells are maintained, and in some cases they may even be stimulated to replenish the lost cells of the tumour following treatment, similar to a wound healing response in normal tissues (Maitland and Collins 2010). Indeed, there are numerous examples in the literature demonstrating that treatment of tumours (colon, breast, glioblastoma) can result in enrichment of the cancer stem cell population and a resultant secondary, more aggressive therapy-resistant tumour (Calcagno et al. 2010; Creighton et al. 2009; Dylla et al. 2008; Phillips et al. 2006). There is therefore the necessity to include a strategy to eliminate cancer stem cells as part of any novel therapeutic aim.

## 4.4 Therapy Resistance Mechanisms and Opportunities for Novel Targets

A variety of proven and potential therapy resistance mechanisms have been described in prostate cancer. Some relate to the tumour as a whole, whilst others are specific to the presence, activity and unique characteristics of cancer stem cells. In addition, the identification of several possible mechanisms of resistance allows us to characterise the failure of a treatment whilst providing indications of novel therapeutic targets. This strategy should have the potential to inhibit the resistance mechanisms and thus sensitise the cancer stem cells to traditional treatments. Examples of these mechanisms and how they might be overcome are described below.

### *4.4.1 Hormone Resistance*

Once advanced prostate cancer becomes hormone resistant, the therapeutic arsenal is diminished. There are several mechanisms of hormone resistance, which primarily relate to an alternative regulation of the androgen/androgen receptor axis (Bluemn and Nelson 2012; Waltering et al. 2012). These include androgen receptor gene amplification, which results in overexpression of the androgen receptor. There is also mutation of the androgen receptor gene, which results in an increased sensitivity to lower levels of androgen or a diversification of the receptor (increased promiscuity) that allows the androgen receptor to respond to and be activated by other hormones such as oestrogen and progesterone. Mutations can also result in constitutive expression of androgen, which then allows crosstalk between pathways. Activation by non-hormone growth factors in a ligand-independent fashion also occurs. Here, the androgen receptor is stimulated by either EGFR (epidermal growth factor receptor) or HER2 (human epidermal growth factor receptor 2), known as the outlaw pathway (Bonkhoff 2012). In addition, androgen receptor isoforms resulting from splice variants and gene rearrangements have been identified that result in expression of receptor variants that lack the ligand binding domain which are constitutively active in the absence of androgens (Dehm et al. 2008; Dehm and Tindall 2011).

Recent work has demonstrated intracrine de novo steroidogenesis following androgen ablation therapy and the onset of castration resistance (Locke et al. 2008; Montgomery et al. 2008). This occurs in the presence of low or negligible levels of circulating (exogenous) androgen and androgens produced *within* tumours at a level that can activate AR-target genes. The bulk of experimental evidence has shown that prostate cancer stem cells are androgen receptor negative and therefore do not respond directly to hormonal stimuli (Collins et al. 2005). However, it is conceivable that the cancer stem cells may respond to signals from androgen-responsive progenitor and stromal cells. It has been shown in cell lines that cells can be androgen responsive without being androgen dependent, so depletion of androgens does not signal the death knell for these cells (Marques et al. 2005).

## *4.4.2 Radioresistance*

Radioresistance is a consideration in all cancers where radiotherapy is a treatment modality. Resistance can arise through a variety of mechanisms, for example, gross mechanisms related to tumour mass and location and more refined molecular mechanisms related to activation of radiation-responsive signalling pathways and also specific mutations that affect the radiation response.

### 4.4.2.1 Role of Tumour Size, Location and Genetic Characteristics

When discussing failure of radiotherapy, there is an initial distinction between non-responsive tumours and radiorecurrent tumours. In some patients, radiotherapy simply does not provide a cure, despite extensive optimisation with the aim of administering carefully controlled doses to maximise impact on tumour and minimise side effects (Brawer 2002). Nonresponsive tumours can arise from inherent radioresistance of the tumour due to their genetic landscape, inaccurate external beam therapy so the whole tumour(s) is not irradiated and the presence of hypoxic regions within the tumour (Bonkhoff 2012). Refinements in imaging techniques are improving the accuracy of both external beam radiotherapy and brachytherapy (e.g. ultrasound-guided brachytherapy) to ensure that the whole tumour is targeted. Hypoxia is a marker for biochemical failure of radiotherapy in prostate cancer, and studies have shown that cells in hypoxic regions undergo a cell cycle arrest and hence reduced apoptosis (Milosevic et al. 2012). Tumours preferentially develop hypoxic regions due to disorganised angiogenesis, anaemia associated with a large tumour burden and increased interstitial pressure (Seiwert et al. 2007).

Hypoxia is thought to contribute to radioresistance because radiation exerts its DNA-damaging effect (at least partly) through free radicals including oxygen radicals (reactive oxygen species (ROS)) (Cook et al. 2004). In addition, breast cancer stem cells have been shown to have increased production of ROS-quenching enzymes, leading to neutralisation of free radicals and reduced DNA damage (Diehn et al. 2009). In a prostate cancer in a cell line model, it was shown that hypoxia treatment resulted in increased colony-forming ability and increased expression of HIF-1α and HIF-2α as well as Nanog and Oct3/4. The cells with higher colony-forming ability had induced CD44 and ABCG2 expression. Therefore, hypoxia treatment resulted in an increase of stem cell-like properties in this study (Ma et al. 2011). The level of hypoxia, be it chronic or acute, affected disease outcome, with acute hypoxia increasing clonogenic activity and chronic hypoxia inducing cell death (Dai et al. 2011).

#### 4.4.2.2 Anti-apoptotic Factors and Radioresistance

Another mechanism employed by cancer stem cells is evasion of apoptosis (Fulda and Pervaiz 2010). Upregulation of Inhibitor of Apoptosis (IAP) family members has been observed in cancer stem cells (Liu et al. 2006). One family member, Survivin, is involved in regulation of normal stem cells and is a known radiation-resistance factor due to interaction with DNA repair factors (Reichert et al. 2011). Survivin is a potential therapeutic target since knockdown or inhibition of Survivin leads to decreased survival. Indeed, Survivin antagonists have been used in clinical trials to target prostate cancer, and so it remains a strategy with potential (Altieri 2012). In addition the Bcl-2 family can have anti-apoptotic properties that lead to radioresistance, and Bcl-2 overexpression is observed and associated with both stem cells and with aggressive prostate cancer (An et al. 2007).

#### 4.4.2.3 Signalling Pathways

Radiation induces activation of a variety of signalling pathways that govern cell survival, growth, proliferation, senescence, invasion, motility and DNA repair (Skvortsova et al. 2008). Understanding these pathways allows us to determine the signalling required for cell death and successful treatment, and which pathways contribute most to radioresistance and therapy failure. This information has been used by the Radiation Therapy Oncology Group to identify biomarkers that can be used to predict treatment failure. The most significant markers are p16 (proliferation), Ki67 (proliferation), MDM2 (degradation of p53 and reduced apoptosis/increased proliferation), COX-2 (pro-inflammatory) and Protein Kinase A (Bonkhoff 2012). In addition, radioresistance can be generated by radiation-induced activation of EGFR through the PI3K/Akt/mTOR pathway, which leads to increased HIF-1$\alpha$ and increased VEGF, which in turn promotes angiogenesis (Shi et al. 2007). Alternatively, loss of PTEN leads to activation of the PI3K/Akt/mTOR pathway, phosphorylation of androgen receptor making it hypersensitive and downregulation of p27, which together result in increased cell proliferation, and is implicated as a further radioresistance mechanism (Heinlein and Chang 2004).

### 4.4.3 *Chemoresistance*

The main chemotherapeutic agent used in prostate cancer is docetaxel, and there have been few advances in chemotherapy for prostate cancer since its introduction. In order to design novel chemotherapies for prostate cancer, one has to consider the potential resistance mechanisms and unintended consequences of new agents.

#### 4.4.3.1 Adaptive Mutations and Selection

Decisions about the optimum doses of various chemotherapies to shrink prostate tumours are made on the basis of maximal tolerated dose (to the patient). Little is known about the effective dose distribution of chemotherapeutics within a tumour mass. Existing evidence from other cancer stem cell systems, and data accumulating about the expression of various drug efflux pump proteins, suggests that the dosages required to achieve toxicity in hypoxic and cancer stem cell-enriched regions of a tumour will be considerably higher when compared to those for a small spheroid grown in normoxic conditions or indeed two-dimensional cultures of cells on plastic substrates (as is most commonly used in preclinical high-throughput screening). Without a treatment combination to prevent drug efflux and to increase tumour permeability, all attempts to achieve toxic doses for cancer stem cells might be fruitless. In addition, the sublethal doses may have a similar effect to persistent treatment with low levels of specific toxins resulting in a resistant phenotype. In which cell therefore do such adaptive mutations arise? Short-term survival could be achieved by mutation in the replicating fraction of the tumour (where mutations most easily arise and can be established). It has been proposed that such changes arise, for example, after hormone therapies, in luminal tumour cells expressing extremely low levels of AR, with no need for a cancer stem cell. If the stem cells and cancer stem cells are AR-, as we propose, then such adaptation is more likely to be achieved in the replicating bulk population. In this case, we do currently have a potential genotypic marker of the phenotype, namely, the amplification of AR seen in about 30% tumour populations after emergence of castration resistance (Haapala et al. 2007). Such amplification is a classical cellular reaction to low substrate concentrations. One would predict that the mutation would not be required in the persisting stem cells, although CRPC is an extremely heterogeneous disease, and there remains the possibility that a stem-like cell can be regenerated (Fig. 4.3a, b(i)) from a mutated progenitor. Intriguing questions remain to be answered, and the definitive proof can only come from human tumours, rather than cell lines, where selective pressure (to grow) has already been applied for many years.

#### 4.4.3.2 Anti-angiogenic Factors Lead to Hypoxia that Stimulates Cancer Stem Cell Enrichment

VEGF-neutralising antibodies have been used in the clinic for a variety of cancers including colorectal cancer and breast carcinoma (bevacizumab) (Kubota 2012; Koutras et al. 2012). However, some serious consequences of using this anti-angiogenic factor to treat tumours have now been realised. The anti-angiogenic activity can lead to hypoxia in the tumours following the reduction in blood vessel formation, and hypoxia can have the unintended effect of enriching for cancer stem cells and unleashing a recurrent, more aggressive tumour (Conley et al. 2012). This highlights the need to take into account the heterogeneity of the tumours and in this case, ignoring the cancer stem cell population means that therapies have the potential to do more harm than good (Wicha 2008).

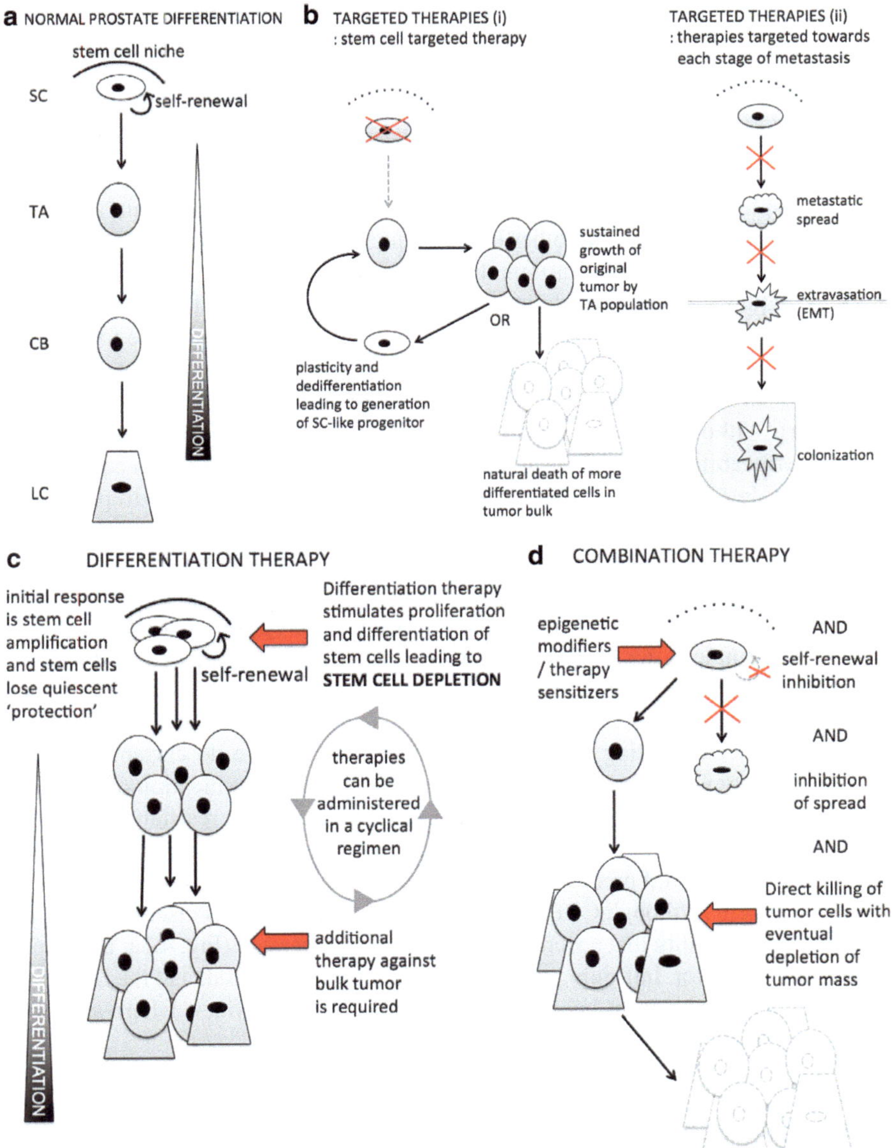


**Fig. 4.3** Clinical applications of cancer stem cell therapies in prostate cancer. (**a**) Normal prostate epithelial cell differentiation initiates from a normal stem cell (SC), which gives rise to a hierarchical cascade of transit-amplifying (TA), committed basal (CB) and terminally differentiated luminal cells (LC). (**b**) Targeted therapies could be used to target cancer stem cell-specific markers (*i*) resulting in reduction of the ability of the tumour to regenerate tumour progenitor cells, and the differentiated cells would gradually die, or there may be dedifferentiation of the progenitor cells to gain stem cell-like features, and the tumour would continue to be propagated. Alternatively, targeted therapies could be designed against each step of metastasis with the intention to inhibit spread of the disease (*ii*). (**c**) Differentiation therapy would have the aim of depleting the cancer stem cell fraction such that more differentiated cells would be produced that are susceptible to traditional therapies. Stimulating the cancer stem cells to proliferate would reduce the protection of their niche. This therapy could be administered cyclically to ensure elimination of all cancer stem cells and all bulk tumours. (**d**) Considering the heterogeneity of prostate tumours, combination therapy is likely to yield greater success. Targeting the stem cells as well as the bulk population and preventing spread of the tumour would be an optimum multipronged approach

#### 4.4.3.3 Signalling Pathways

The growth of bulk tumours in the prostate is known to be dependent on androgenic stimuli. Salvage pathways can provide some alternative sources of hormone, but the tumours can still develop resistance to new generation treatments such as abiraterone (Li et al. 2012a). There are two explanations for such resistance: either the cancer cells have found yet another alternative source of hormone, or they have become truly androgen independent as a means of achieving the same growth-related signalling to that stimulated by the hormones. One commonly explored pathway, which has been shown to intersect with AR signalling, is the response to pro-inflammatory cytokines such as IL-1, IL-4 (Lee et al. 2003, 2008, 2009) and IL-6 (Azevedo et al. 2011). Since cancer stem cells in prostate already display such an inflammation-related phenotype, for example, expressing high levels of IL-6 and IL-6R, as well as phosphorylated STAT3, and NFκB, they already have the potential to respond in this way (Birnie et al. 2008). IL-4 contributes to survival of colon cancer stem cells conferring resistance to chemotherapy, partly through induction of the anti-apoptotic regulator, Survivin (Di Stefano et al. 2010). It is feasible that the cytokines regulate prostate cancer stem cells in a similar manner. Indeed, NFκB is highly expressed in prostate cancer stem cells and progenitor cells and when inhibited by parthenolide resulted in apoptosis of the cancer stem cells (Birnie et al. 2008). However, generalised NFκB inhibition can result in host toxicity (Aggarwal and Sung 2011). We, and others, have proposed the link between persistent inflammation, previously established on epidemiological grounds and with animal models, and prostate cancer induction in vivo (Maitland and Collins 2008; De Marzo et al. 2007).

Other growth factor-linked pathways such as those for Notch, Wnt and Hedgehog signalling have also been shown to be active in cancer stem cells from prostate and other human tumours (Takebe et al. 2011). The components of the pathways have therefore been considered to be therapeutic targets. Wnt signalling has been associated with self-renewal in normal stem cells, but it has also been linked to cancer and cancer stem cells. In terms of stem cell control, along with self-renewal, there is also controlled proliferation, inhibition of differentiation, maintenance and survival (Reya and Clevers 2005). It is likely that these pathways, whilst acting to give cancer stem cells similar characteristics as normal stem cells, e.g. self-renewal, in the cancer phenotype, the signalling is dysregulated. Activation of the Hedgehog pathway induces cell proliferation, and inhibition of the pathway inhibits growth (Sanchez et al. 2004). Hedgehog is active in prostate cancer and continuous activation allows a progression both from normal to tumorigenic cells and from tumorigenic to invasive cells (Karhadkar et al. 2004). Similarly, Notch signalling has a broad spectrum of activities including survival, differentiation and proliferation, and Notch-1 is overexpressed in metastatic prostate cancer (Wang et al. 2010). Inhibition of Notch signalling is therefore another valid therapeutic approach, and indeed there is evidence that Notch inhibition leads to reduced growth, migration and invasion and increased apoptosis (Wang et al. 2010). However, tests on all these pathways are often extensively characterised in cell lines, where the response can be neat and unambiguous. There is still the need for the progression to patient samples before embarking on clinical trials. Since such 'embryonal' signalling pathways are essential for cell survival, there is the potential for crosstalk and redundancy between pathways or even bypass pathways that will cancel out the effect of inhibitors. However, some agents are now being tested in clinical trials, and there is an expectation that they will succeed in targeting cancer stem cells (Takebe et al. 2011).

All such pathways present very real barriers to therapeutic development, as they are essential for maintenance of normal tissue architecture and in some cases are necessary for stem cell survival. In addition, there is extensive crosstalk between pathways and unpredictable side effects. Unless a particular signalling intermediate is uniquely expressed in the cancer stem cells, the potential for short- and more seriously long-term side effects, following inhibition of these pathways, in normal tissues and normal stem cells is very real.

#### 4.4.3.4 Flexibility of Metabolic Status

There is now evidence that cancer cells have a different metabolism compared to normal cells, in particular relating to glucose metabolism and oxidative metabolism. Perhaps more significantly, glioma stem cells have also been shown to switch their primary metabolism pathways (Vlashi et al. 2011). This chameleon property of these cells contributes to the complexity of developing novel therapies. Interestingly, one recent study showed a relationship between CD44 expression and glucose metabolism (Tamada et al. 2012). CD44 is a marker of basal/progenitor prostate epithelial cells, and expression is associated with tumour-initiating properties to many cell types including prostate (Maitland et al. 2011). Cancer cells typically produce energy using glycolysis in preference to mitochondrial respiration, which results in reduced reactive oxygen species (ROS) and therefore resistance to ROS-inducing therapies, e.g. radiation and some chemotherapeutic drugs. CD44 interacts with a component of the glycolysis pathway such that it is favoured over respiration in p53-deficient and hypoxic cancer cells. Furthermore, when CD44 was inhibited using siRNA, this resulted in a shift of metabolism and also sensitised the cells to cisplatin, thus linking a progenitor phenotype and stem cell marker with a known cancer phenotype (the Warburg effect). However, this study was carried out only in cell lines under normoxic conditions. It would be most intriguing to characterise these relationships in primary cultures and real patient samples.

## 4.5 Immune Evasion of Stem Cells

The gold standard for cancer stem cell phenotype analysis remains their ability to reinitiate tumour growth in immunocompromised mice (Brunner et al. 2012). The power of the immune system to eliminate even the cancer stem cell population is demonstrated by the increasing tumour take rates (A. Collins, manuscript in preparation) when the degree of immunodeficiency is increased from nude (athymic so deficient in CD4+ and CD8+ T cells), NOD-SCID (impaired ability to make T or B lymphocytes or activate some components of the complement system), to the $RAG2^{-/-}$, gamma $c^{-/-}$ mouse which totally lacks B and T cells, but also an essential natural killer (NK) cell response. The expression of NK targets on cancer stem cells has already been noted in other cell systems (Jewett and Tseng 2012; Jewett et al. 2012), most notably in melanoma (Pietra et al. 2009). This therefore poses the question about how cancer stem cells can function in the immunocompetent environment in a human patient, an issue which has been summarised by Qi et al. (2012). It is likely that the cancer stem cells can induce an immune tolerance, in addition to the interaction and cooperation by the stem cell niche and the tumour microenvironment. Such immune evasion could conceivably be overcome, perhaps even allowing a vaccination strategy, but again necessarily as part of a combination therapy.

## 4.6 Tumour Dormancy and Stem Cell Quiescence

Relapse of a cancer after treatment can take months or sometimes years. The period where there is no apparent tumour growth is termed dormancy (Aguirre-Ghiso 2007). Dormancy can occur due to internal factors such as a block in proliferation and a cell cycle arrest related to cell signalling, external pressures within the microenvironment (Bragado et al. 2012; Sottocornola and Lo Celso 2012) such as a lack of angiogenesis or activation of the host immune system, or induction of apoptosis. It is thought that cellular quiescence is one of the internal mechanisms by which the cells are arrested. Quiescence is considered likely to contribute to tumour dormancy because it has the potential to be reversed as a result of a change in external signals and the microenvironment. For this reason, the more permanent and less reversible senescence is less likely to be involved in dormancy but has been acknowledged as a tumour-suppressive mechanism (Aguirre-Ghiso 2007).

In terms of the therapeutic opportunity with dormancy, if we were able to *maintain* a dormant tumour or dormant metastases, then the patient could live with asymptomatic minimal residual disease. Maintenance, rather than elimination, in this case could be an acceptable outcome. Indeed, maintenance therapy is in routine use for myeloma and lymphoma, albeit with continual room for improvement (Maiolino et al. 2012; Badros 2010), but not yet for prostate cancer. Of course in terms of therapy, this would require more understanding of the trigger that activates cells to exit dormancy. This could be due to different growth-promoting signals or else an adaptation of the cells, which allows them to respond differently to growth-promoting signals. Significantly, once relapse has occurred and primary tumour or metastatic growth has been initiated, the cancer presents a difficult therapeutic challenge.

It is also possible to marry tumour dormancy with cancer stem cells (Kusumbe and Bapat 2009). There is now evidence that cancer stem cells are predominantly, though not entirely, quiescent. Therefore, it has been postulated that quiescent cancer stem cells are responsible for preserving the potential for recurrent primary and metastatic tumours (Moore et al. 2012; Moore and Lyle 2011). Naturally, quiescent cells are not sensitive to chemotherapy and radiotherapy that predominantly target proliferating cells, and so alternative therapies are required for this subpopulation. Evidence of quiescent cancer stem cells has been presented for melanoma, pancreatic, breast, ovarian and haematopoietic malignancies (Buczacki et al. 2011). One study in mouse prostate cites TGF-β as a key factor in maintaining normal prostate stem cell dormancy (Salm et al. 2005). Assays used to identify quiescent cells include staining with the Ki67 proliferation marker where Ki67-negative cells being indicative of the G0 stage of the cell cycle. Also, label-retaining experiments have been used with the principle that a label which inserts into the cell membrane (e.g. PKH26) is diluted as cells divide and proliferate whilst being retained in nondividing or quiescent cells (Moore and Lyle 2011). This has been shown in a variety of tumour types (Chen et al. 2012; Munoz et al. 2012; Li et al. 2012b), and indeed it has even been suggested that these quiescent cells are more invasive and more aggressive, at least in breast cancer (Pece et al. 2010). If this is the case in prostate cancer, then quiescent cancer stem cells can be thought of as 'the sleeping assassin'.

## 4.7 Functional Assays for Cancer Stem Cell Therapeutics

### *4.7.1 Transplantation*

Although the gold standard for identifying cancer stem cells has been tumour initiation in immunodeficient mice (Brunner et al. 2012), this has been brought into question. In a melanoma model, it was shown that tumour initiation was not a feature of a *rare* population of cells and in fact almost 33% of cells had the ability to form tumours. However, efficiency of tumour formation depended greatly on the genetic background and therefore the permissiveness of the mouse model (Quintana et al. 2008). This brought into question the core of the cancer stem cell hypothesis, i.e. that not all cells are equal. However, on testing in pancreatic, head and neck and lung models, Ishizawa et al. (2010) showed that whilst there was variation between mouse models, in these tumours, there remained a rare subpopulation of tumour-initiating cells. Currently, there is nothing to replace the use of in vivo models to test novel therapeutics and to assess the enrichment of stem cells as well as testing serial passaging of human tumours. This latter assay can distinguish between cells that can initiate primary tumours but are unable to initiate secondary tumours, from true cancer stem cells which can initiate tumours potentially indefinitely over multiple in vivo passages (Brunner et al. 2012).

### *4.7.2 Lineage Tracking*

A recent article by Wright (2012) has argued against the use of mouse models and indeed against the selection of subpopulations of cells in vitro to identify stem cells and cancer stem cells. In contrast, the use of lineage tracking in vitro and in vivo to determine the *fate* of cells rather than their surface marker expression at a single point in time is proposed. The most elegant studies of lineage tracking have been demonstrated in the small intestine from Hans Clevers' group, and these indeed have allowed identification of stem cells as well as elucidation of their life cycle (Snippert and Clevers 2011). Whilst not disagreeing with this viewpoint, it is not entirely practical to do in vivo lineage tracking for all tissues. However, within the prostate, lineage tracking in culture has been attempted, and indeed infection of primary prostate epithelial cells with lentivirus encoding fluorescent marker genes under the control of differentiation-specific promoters has been undertaken. This showed proof-of-principle results to indicate a method of tracking differentiation (Frame et al. 2010). Any such lineage tracking firstly has to overcome limitations of the vector as well optimising the strategy for the specific cell types. Other methods of lineage tracking can include mapping genetic changes in human cancers and evolution of tumours through clonal selection, most elegantly done in leukaemia (Ding et al. 2012). Alternatively, in prostate there has been a lineage tracking study of prostate stem cells within benign tissue (Blackwood et al. 2011), which used laser capture microdissection to assess respiratory chain defects that resulted from mitochondrial DNA mutations. This allowed analysis of individual acini within the prostate gland and identification of a single progenitor cell.

### 4.7.3 *Stem Cell Markers*

Markers of stem cells include surface markers that define a rare subpopulation of cells with the characteristics of stem cells including self-renewal and differentiation. Cells can be selected using specific antibodies to these markers and FACs sorting or MACs sorting (Visvader and Lindeman 2008; Miki and Rhim 2008). In addition, isolation of side populations has been used as another method to identify and extract stem cell populations. This approach is based on the principle that stem cells have increased drug efflux proteins including aldehyde dehydrogenase (ALDH) (Ginestier et al. 2007; Ma and Allan 2011) and ABC transporters (Dean et al. 2005; Ding et al. 2010; Elliott et al. 2010; Scotto 2003; Tanei et al. 2009).

### 4.7.4 *In Vitro Assays*

There are a variety of in vitro assays to test for stem cells and cancer stem cells. Typically, these assays are focussed on examining the *potential* of the cells. Clonogenic assays assess the ability of the cell to give rise to new progenitor cells, and importantly secondary colony-forming assays can distinguish between highly proliferating progenitor cells and cells with stem cell properties. Also, cells with stem cell-like features are more able to form 3D spheres in sphere-forming assays. Other assays in vitro include proliferation assays to determine when the cells reach exhaustion as well as differentiation assays by adding different media and growth factors, or in the case of prostate adding serum, stroma and DHT can differentiate the cells (Swift et al. 2010). Differentiation can be detected using immuno-staining for basal and luminal cytokeratins, as well as differentiation markers, such as androgen receptor, prostatic acid phosphatase (PAP) and PSA. Another in vitro assay is the label-retaining assay that has been described above (Moore and Lyle 2011).

## 4.8 Clinical Applications of Cancer Stem Cell Therapies in Prostate Cancer

### *4.8.1 The Role of Stem Cells in Multifocal Prostate Cancer*

We have established that design of novel therapeutics for prostate cancer must take into account the cancer stem cell population. Alongside this consideration, we have to evaluate the heterogeneous tumour as a whole, which has resulted from a series of mutations resulting in clonal selection, evolution and expansion (Marusyk et al. 2012). The consequences of these events and the potential for more than one cell of origin with an ability to initiate tumours can be multifocal disease, typically in two thirds of patients. Multifocal disease is associated with more aggressive, higher-grade prostate tumours than unifocal disease (Djavan et al. 1999). In therapeutic terms, there is also the consideration of adaptation of the tumour to promote invasion and metastasis (Fig. 4.3b(ii)). Heterogeneity may also exist at this stage within the cancer stem cell population, as there is likely to be considerable plasticity of the stem cell phenotype, given the different properties of the tumours resulting from different selection pressures due to external signalling, tumour microenvironment and stem cell niche (van der Pluijm 2011; Risbridger and Taylor 2008; Kelly and Yin 2008).

### *4.8.2 Combination Therapies*

Monotherapies can be successful depending on the tumour type and the mode of action. However, with the huge variety of cancers and the potential adaptation and mutation in response to environmental selection pressures and indeed treatment itself, along with increased understanding of the mechanisms of therapy resistance, combination therapies are likely to be effective and longer lasting (Fig. 4.3d). Recent proposals to combine next-generation anti-androgen response modifiers, such as MDV3100 and abiraterone, may fail to exploit the biggest advantage of combination therapies, i.e. the ability to block different or complementary pathways.

In prostate cancer, radiation alone can lead to recurrence, hormone therapy leads to resistance, and responsiveness to chemotherapy is short-lived. Along with these observations and known heterogeneity of tumours, combination therapies are predicted to be a more successful strategy in the future of prostate cancer treatment, and indeed it has already been embraced. As described above, radiotherapy is not 100% effective due to inherent tumour resistance and to dose limits to reduce toxicity. Therefore, combining radiotherapy with other treatment modalities could greatly enhance therapeutic effect. This is currently already carried out in terms of using radiotherapy alongside surgery or radiotherapy with androgen ablation treatments. However, more targeted approaches are being investigated that could enhance radiosensitivity of prostate tumour cells.

Inhibitors of poly (ADP ribose) polymerase (PARP) are under consideration as a treatment for prostate cancer and are being used in current clinical trials. This enzyme is essential for the repair of single-strand DNA breaks and can complement radiotherapy. PARP inhibitor use for prostate tumours has some promise because a proportion of prostate cancer cells are mutated in the BRCA1 and BRCA2 genes and also in PTEN; these mutations indicate a defect in homologous recombination, and inhibition of other repair pathways with PARP inhibitors acts as a double hit to the cancer cell (de Bono et al. 2011). There is also evidence that PARP inhibitors inhibit ETS gene-driven prostate cancer models (Brenner et al. 2011).

Finally, several other therapeutics are being considered as partners for radiotherapy, reviewed in (Verheij et al. 2010), such as EGFR inhibitors, anti-angiogenic drugs and apoptosis-modulating agents.

### *4.8.3 Differentiation Therapies*

The principle of differentiation therapy is to first stimulate stem cells out of quiescence or stress-induced growth arrest, thereby inducing proliferation and producing differentiated cycling progeny (Sell 2004). Such ‘pre’-treatment should result in the cells being susceptible to radiation and anti-proliferative drugs as well as drugs that target more differentiated cells. Ultimately, the aim will be to deplete the cancer stem cell stores to reduce the regeneration and recurrence potential of the tumour (Fig. 4.3c). One note of caution when considering differentiation therapy is that normal stem cells should not be affected; otherwise, they too could be depleted with catastrophic consequences for the patient as a whole (Moore and Lyle 2011). One example of a therapy which has had the desired effect of reducing cancer stem cell frequency is use of a DLL4 inhibitor (Hoey et al. 2009). DLL4 is a potent ligand for the Notch pathway and when inhibited, there was a concomitant reduction in tumour recurrence and angiogenesis, as demonstrated in a colon cancer xenograft model. However, similar untargeted DLL4 inhibition/knockout in a mouse model had serious longer-term consequences on vascular functions, resulting in multiple liver haemangiomas (Yan et al. 2010). The observation that the haemangiomas arose as a long-term (8 week) consequence of chronic DLL4 block implies that side effects of cancer stem cell therapies may only emerge in human patients after proportionately long time scales and could be managed in advanced cancer patients for whom the short-term prognosis is poor.

In terms of clinical rather than experimental success, retinoids as the differentiation stimulator in acute promyelocytic leukaemia have improved patient survival, with the application of ATRA (all-trans-retinoic acid) leading to release of the differentiation block that occurs as a result of the PML-RARα fusion (Cruz and Matushansky 2012). There has been less success in solid tumours due to a poorer understanding of differentiation pathways and perhaps the inability of the cancer cells to differentiate, depending on their defining mutations and the likelihood that many pathways are affected (Rane et al. 2012).

Epigenetic modifiers including histone deacetylase (HDAC) inhibitors and DNA methyltransferase (DNMT) inhibitors were initially studied due to their ability to promote differentiation, which was associated with reduced cell proliferation (Piekarz and Bates 2009). It is now known that the differentiation observed was due to perturbations of DNA methylation and histone acetylation. This feature was clearly of interest for treatment of cancer cells, and indeed monotherapies of DNMT inhibitors and HDAC inhibitors have been used against AML (acute myelogenous leukaemia) and T cell lymphomas. However, in solid tumours, it is thought that they will be most effective as part of a combination therapy (Pili et al. 2012). Proposed therapies would combine the differentiation-inducing ability of the epigenetic modifiers with toxic drugs or radiation. In addition, they have the effect of decondensing the chromatin, such that the DNA is more susceptible to damage by the latter agents.

### 4.8.4 Targeted Therapies

In order to design therapies that target cancer stem cells specifically, it is possible to design immunotherapies or small molecule inhibitors that target stem cell-specific surface markers (Fig. 4.3b(i)) (Qi et al. 2012). The ability to select stem cells from tumours has allowed extensive microarray and proteomic studies to identify cancer stem cell-specific markers. The most crucial criteria when choosing such a target are that it should be tissue specific and should not be expressed on normal stem cells. The successful targeting molecule should necessarily be required for cancer stem cell survival, to reduce the chances of gene expression silencing as an immediate resistance mechanism. Along with targeting stem cells, targeted therapy should be designed against the other steps of tumour progression including invasion and metastasis (Fig. 4.3b(ii)).

## 4.9 Conclusions

We have provided here an overview of the complexity of prostate cancer treatment, taking into account the progression of the disease, current treatments and the problem of therapy resistance, as seen from a cancer stem cell perspective. However, the mechanisms of therapy resistance are many and varied. In order to improve prostate cancer treatment, an understanding of the molecular mechanisms of resistance as well as the biology of the disease is required. Despite many billions of dollars expended on 'new' treatments, and thousands of publications in the scientific literature, we still measure advances in months rather than genuinely increased survival for advanced prostate cancer. Adoption of a stem cell-based approach to new therapy targets to complement the improved tolerability of the new therapies could provide the paradigm shift in therapeutic outcome which researchers seek.

**Acknowledgements** Our research in prostate cancer stem cells was supported by programme and project grants from Yorkshire Cancer Research (Registered Charity 516898).

## References

Aggarwal BB, Sung B (2011) NF-kappaB in cancer: a matter of life and death. Cancer Discov 1:469–471

Aguirre-Ghiso JA (2007) Models, mechanisms and clinical evidence for cancer dormancy. Nat Rev Cancer 7:834–846

Altieri DC (2012) Targeting survivin in cancer. Canc Lett. http://dx.doi.org/10.1016/j.conlet.2012.03.005

An J, Chervin AS, Nie A, Ducoff HS, Huang Z (2007) Overcoming the radioresistance of prostate cancer cells with a novel Bcl-2 inhibitor. Oncogene 26:652–661

Azevedo A, Cunha V, Teixeira AL, Medeiros R (2011) IL-6/IL-6R as a potential key signaling pathway in prostate cancer development. World J Clin Oncol 2:384–396

Badros AZ (2010) The role of maintenance therapy in the treatment of multiple myeloma. J Natl Compr Canc Netw 8(Suppl 1):S21–27

Bastus NC, Boyd LK, Mao X, Stankiewicz E, Kudahetti SC et al (2010) Androgen-induced TMPRSS2:ERG fusion in nonmalignant prostate epithelial cells. Cancer Res 70:9544–9548

Birnie R, Bryce SD, Roome C, Dussupt V, Droop A et al (2008) Gene expression profiling of human prostate cancer stem cells reveals a pro-inflammatory phenotype and the importance of extracellular matrix interactions. Genome Biol 9:R83

Blackwood JK, Williamson SC, Greaves LC, Wilson L, Rigas AC et al (2011) In situ lineage tracking of human prostatic epithelial stem cell fate reveals a common clonal origin for basal and luminal cells. J Pathol 225:181–188

Bluemn EG, Nelson PS (2012) The androgen/androgen receptor axis in prostate cancer. Curr Opin Oncol 24:251–257

Bonkhoff H (2012) Factors implicated in radiation therapy failure and radiosensitization of prostate cancer. Prostate Canc 2012: 593241

Bragado P, Sosa MS, Keely P, Condeelis J, Aguirre-Ghiso JA (2012) Microenvironments dictating tumor cell dormancy. Recent Results Cancer Res (Fortschritte der Krebsforschung Progres dans les recherches sur le cancer) 195:25–39

Brawer MK (2002) Radiation therapy failure in prostate cancer patients: risk factors and methods of detection. Rev Urol 4(Suppl 2):S2–S11

Brenner JC, Ateeq B, Li Y, Yocum AK, Cao Q et al (2011) Mechanistic rationale for inhibition of poly(ADP-ribose) polymerase in ETS gene fusion-positive prostate cancer. Cancer Cell 19:664–678
Brunner TB, Kunz-Schughart LA, Grosse-Gehling P, Baumann M (2012) Cancer stem cells as a predictive factor in radiotherapy. Semin Radiat Oncol 22:151–174
Buczacki S, Davies RJ, Winton DJ (2011) Stem cells, quiescence and rectal carcinoma: an unexplored relationship and potential therapeutic target. Br J Cancer 105:1253–1259
Calcagno AM, Salcido CD, Gillet JP, Wu CP, Fostel JM et al (2010) Prolonged drug selection of breast cancer cells and enrichment of cancer stem cell characteristics. J Natl Cancer Inst 102:1637–1652
Chen Y, Li D, Wang D, Liu X, Yin N et al (2012) Quiescence and attenuated DNA damage response promote survival of esophageal cancer stem cells. J Cell Biochem 113(12): 3643–3652
Cheng L, Bostwick DG, Li G, Wang Q, Hu N et al (1999) Allelic imbalance in the clonal evolution of prostate carcinoma. Cancer 85:2017–2022
Collins AT, Berry PA, Hyde C, Stower MJ, Maitland NJ (2005) Prospective identification of tumorigenic prostate cancer stem cells. Cancer Res 65:10946–10951
Conley SJ, Gheordunescu E, Kakarala P, Newman B, Korkaya H et al (2012) Antiangiogenic agents increase breast cancer stem cells via the generation of tumor hypoxia. Proc Natl Acad Sci USA 109:2784–2789
Cook JA, Gius D, Wink DA, Krishna MC, Russo A et al (2004) Oxidative stress, redox, and the tumor microenvironment. Semin Radiat Oncol 14:259–266
Creighton CJ, Li X, Landis M, Dixon JM, Neumeister VM et al (2009) Residual breast cancers after conventional therapy display mesenchymal as well as tumor-initiating features. Proc Natl Acad Sci USA 106:13820–13825
Cruz FD, Matushansky I (2012) Solid tumor differentiation therapy - is it possible? Oncotarget 3:559–567
Dai Y, Bae K, Siemann DW (2011) Impact of hypoxia on the metastatic potential of human prostate cancer cells. Int J Radiat Oncol Biol Phys 81:521–528
De Marzo AM, Meeker AK, Epstein JI, Coffey DS (1998) Prostate stem cell compartments: expression of the cell cycle inhibitor p27Kip1 in normal, hyperplastic, and neoplastic cells. Am J Pathol 153:911–919
De Marzo AM, Platz EA, Sutcliffe S, Xu J, Gronberg H et al (2007) Inflammation in prostate carcinogenesis. Nat Rev Cancer 7:256–269
Dean M, Fojo T, Bates S (2005) Tumour stem cells and drug resistance. Nat Rev Cancer 5:275–284
Dehm SM, Tindall DJ (2011) Alternatively spliced androgen receptor variants. Endocr Relat Cancer 18:R183–196
Dehm SM, Schmidt LJ, Heemers HV, Vessella RL, Tindall DJ (2008) Splicing of a novel androgen receptor exon generates a constitutively active androgen receptor that mediates prostate cancer therapy resistance. Cancer Res 68:5469–5477
Di Stefano AB, Iovino F, Lombardo Y, Eterno V, Hoger T et al (2010) Survivin is regulated by interleukin-4 in colon cancer stem cells. J Cell Physiol 225:555–561
Diehn M, Cho RW, Lobo NA, Kalisky T, Dorie MJ et al (2009) Association of reactive oxygen species levels and radioresistance in cancer stem cells. Nature 458:780–783
Ding XW, Wu JH, Jiang CP (2010) ABCG2: a potential marker of stem cells and novel target in stem cell and cancer therapy. Life Sci 86:631–637
Ding L, Ley TJ, Larson DE, Miller CA, Koboldt DC et al (2012) Clonal evolution in relapsed acute myeloid leukaemia revealed by whole-genome sequencing. Nature 481:506–510
Djavan B, Susani M, Bursa B, Basharkhah A, Simak R et al (1999) Predictability and significance of multifocal prostate cancer in the radical prostatectomy specimen. Tech Urol 5:139–142
Drachenberg DE (2000) Treatment of prostate cancer: watchful waiting, radical prostatectomy, and cryoablation. Semin Surg Oncol 18:37–44
Dylla SJ, Beviglia L, Park IK, Chartier C, Raval J et al (2008) Colorectal cancer stem cells are enriched in xenogeneic tumors following chemotherapy. PLoS One 3:e2428
Eastham JA, Stapleton AM, Gousse AE, Timme TL, Yang G et al (1995) Association of p53 mutations with metastatic prostate cancer. Clin Cancer Res 1:1111–1118

Elliott A, Adams J, Al-Hajj M (2010) The ABCs of cancer stem cell drug resistance. IDrugs 13:632–635

Etzioni R, Gulati R, Tsodikov A, Wever EM, Penson DF et al (2012) The prostate cancer conundrum revisited: treatment changes and prostate cancer mortality declines. Cancer 118(23):5955–5963

Frame FM, Hager S, Pellacani D, Stower MJ, Walker HF et al (2010) Development and limitations of lentivirus vectors as tools for tracking differentiation in prostate epithelial cells. Exp Cell Res 316:3161–3171

Fulda S, Pervaiz S (2010) Apoptosis signaling in cancer stem cells. Int J Biochem Cell Biol 42:31–38

Germann M, Wetterwald A, Guzman-Ramirez N, van der Pluijm G, Culig Z et al (2012) Stem-like cells with luminal progenitor phenotype survive castration in human prostate cancer. Stem Cells 30:1076–1086

Ginestier C, Hur MH, Charafe-Jauffret E, Monville F, Dutcher J et al (2007) ALDH1 is a marker of normal and malignant human mammary stem cells and a predictor of poor clinical outcome. Cell Stem Cell 1:555–567

Goldstein AS, Huang J, Guo C, Garraway IP, Witte ON (2010a) Identification of a cell of origin for human prostate cancer. Science 329:568–571

Goldstein AS, Stoyanova T, Witte ON (2010b) Primitive origins of prostate cancer: in vivo evidence for prostate-regenerating cells and prostate cancer-initiating cells. Mol Oncol 4:385–396

Haapala K, Kuukasjarvi T, Hyytinen E, Rantala I, Helin HJ et al (2007) Androgen receptor amplification is associated with increased cell proliferation in prostate cancer. Hum Pathol 38:474–478

Haffner MC, Aryee MJ, Toubaji A, Esopi DM, Albadine R et al (2010) Androgen-induced TOP2B-mediated double-strand breaks and prostate cancer gene rearrangements. Nat Genet 42:668–675

Hay CW, McEwan IJ (2012) The impact of point mutations in the human androgen receptor: classification of mutations on the basis of transcriptional activity. PLoS One 7:e32514

Heinlein CA, Chang C (2004) Androgen receptor in prostate cancer. Endocr Rev 25:276–308

Hoey T, Yen WC, Axelrod F, Basi J, Donigian L et al (2009) DLL4 blockade inhibits tumor growth and reduces tumor-initiating cell frequency. Cell Stem Cell 5:168–177

Hudson DL, Guy AT, Fry P, O'Hare MJ, Watt FM et al (2001) Epithelial cell differentiation pathways in the human prostate: identification of intermediate phenotypes by keratin expression. J Histochem Cytochem 49:271–278

Isaacs JT, Coffey DS (1981) Adaptation versus selection as the mechanism responsible for the relapse of prostatic cancer to androgen ablation therapy as studied in the Dunning R-3327-H adenocarcinoma. Cancer Res 41:5070–5075

Isaacs JT, Coffey DS (1989) Etiology and disease process of benign prostatic hyperplasia. Prostate Suppl 2:33–50

Ishizawa K, Rasheed ZA, Karisch R, Wang Q, Kowalski J et al (2010) Tumor-initiating cells are rare in many human tumors. Cell Stem Cell 7:279–282

Ishkanian AS, Zafarana G, Thoms J, Bristow RG (2010) Array CGH as a potential predictor of radiocurability in intermediate risk prostate cancer. Acta Oncologica 49:888–894

Jewett A, Tseng HC (2012) Potential rescue, survival and differentiation of cancer stem cells and primary non-transformed stem cells by monocyte-induced split anergy in natural killer cells. Cancer Immunol Immunother 61:265–274

Jewett A, Tseng HC, Arasteh A, Saadat S, Christensen RE et al (2012) Natural killer cells preferentially target cancer stem cells; role of monocytes in protection against NK cell mediated lysis of cancer stem cells. Curr Drug Deliv 9:5–16

Jones JS (2011) Radiorecurrent prostate cancer: an emerging and largely mismanaged epidemic. Eur Urol 60(3):411–412

Kallioniemi OP, Visakorpi T (1996) Genetic basis and clonal evolution of human prostate cancer. Adv Cancer Res 68:225–255

Karhadkar SS, Bova GS, Abdallah N, Dhara S, Gardner D et al (2004) Hedgehog signalling in prostate regeneration, neoplasia and metastasis. Nature 431:707–712

Kelly K, Yin JJ (2008) Prostate cancer and metastasis initiating stem cells. Cell Res 18:528–537
Kirby M, Hirst C, Crawford ED (2011) Characterising the castration-resistant prostate cancer population: a systematic review. Int J Clin Pract 65:1180–1192
Koivisto P, Kononen J, Palmberg C, Tammela T, Hyytinen E et al (1997) Androgen receptor gene amplification: a possible molecular mechanism for androgen deprivation therapy failure in prostate cancer. Cancer Res 57:314–319
Koutras AK, Starakis I, Lymperatou D, Kalofonos HP (2012) Angiogenesis as a therapeutic target in breast cancer. Mini Rev Med Chem 12(12):1230–1238
Kubota Y (2012) Tumor angiogenesis and anti-angiogenic therapy. Keio J Med 61:47–56
Kumar A, White TA, MacKenzie AP, Clegg N, Lee C et al (2011) Exome sequencing identifies a spectrum of mutation frequencies in advanced and lethal prostate cancers. Proc Natl Acad Sci USA 108:17087–17092
Kusumbe AP, Bapat SA (2009) Cancer stem cells and aneuploid populations within developing tumors are the major determinants of tumor dormancy. Cancer Res 69:9245–9253
Kyprianou N, Isaacs JT (1988) Activation of programmed cell death in the rat ventral prostate after castration. Endocrinology 122:552–562
Lapidot T, Sirard C, Vormoor J, Murdoch B, Hoang T et al (1994) A cell initiating human acute myeloid leukaemia after transplantation into SCID mice. Nature 367:645–648
Lawson DA, Zong Y, Memarzadeh S, Xin L, Huang J et al (2010) Basal epithelial stem cells are efficient targets for prostate cancer initiation. Proc Natl Acad Sci USA 107:2610–2615
Lee SO, Lou W, Hou M, Onate SA, Gao AC (2003) Interleukin-4 enhances prostate-specific antigen expression by activation of the androgen receptor and Akt pathway. Oncogene 22:7981–7988
Lee SO, Pinder E, Chun JY, Lou W, Sun M et al (2008) Interleukin-4 stimulates androgen-independent growth in LNCaP human prostate cancer cells. Prostate 68:85–91
Lee SO, Chun JY, Nadiminty N, Lou W, Feng S et al (2009) Interleukin-4 activates androgen receptor through CBP/p300. Prostate 69:126–132
Li R, Evaul K, Sharma KK, Chang KH, Yoshimoto J et al (2012a) Abiraterone inhibits 3beta-hydroxysteroid dehydrogenase: a rationale for increasing drug exposure in castration-resistant prostate cancer. Clin Cancer Res 18:3571–3579
Li L, Li B, Shao J, Wang X (2012b) Chemotherapy sorting can be used to identify cancer stem cell populations. Mol Biol Rep 39(11):9955–9963
Limas C, Frizelle SP (1994) Proliferative activity in benign and neoplastic prostatic epithelium. J Pathol 174:201–208
Liu G, Yuan X, Zeng Z, Tunici P, Ng H et al (2006) Analysis of gene expression and chemoresistance of CD133+ cancer stem cells in glioblastoma. Mol Cancer 5:67
Locke JA, Guns ES, Lubik AA, Adomat HH, Hendy SC et al (2008) Androgen levels increase by intratumoral de novo steroidogenesis during progression of castration-resistant prostate cancer. Cancer Res 68:6407–6415
Ma I, Allan AL (2011) The Role of Human Aldehyde Dehydrogenase in Normal and Cancer Stem Cells. Stem Cell Rev 7(2):292–306
Ma Y, Liang D, Liu J, Axcrona K, Kvalheim G et al (2011) Prostate cancer cell lines under hypoxia exhibit greater stem-like properties. PLoS One 6:e29170
Maiolino A, Hungria VT, Garnica M, Oliveira-Duarte G, Oliveira LC et al (2012) Thalidomide plus dexamethasone as a maintenance therapy after autologous hematopoietic stem cell transplantation improves progression-free survival in multiple myeloma. Am J Hematol 87(10):948–952
Maitland NJ, Collins A (2005) A tumour stem cell hypothesis for the origins of prostate cancer. BJU Int 96:1219–1223
Maitland NJ, Collins AT (2008) Inflammation as the primary aetiological agent of human prostate cancer: a stem cell connection? J Cell Biochem 105:931–939
Maitland NJ, Collins AT (2010) Cancer stem cells— a therapeutic target? Curr Opin Mol Ther 12:662–673

Maitland NJ, Frame FM, Polson ES, Lewis LJ, Collins AC (2011) Prostate cancer stem cells: do they have a basal or luminal phenotype? Horm Cancer 2(1):47–61
Mani RS, Tomlins SA, Callahan K, Ghosh A, Nyati MK et al (2009) Induced chromosomal proximity and gene fusions in prostate cancer. Science 326:1230
Marcu M, Radu E, Sajin M (2010) Neuroendocrine transdifferentiation of prostate carcinoma cells and its prognostic significance. Rom J Morphol Embryol 51:7–12
Marin-Aguilera M, Codony-Servat J, Kalko SG, Fernandez PL, Bermudo R et al (2012) Identification of docetaxel resistance genes in castration-resistant prostate cancer. Mol Cancer Ther 11:329–339
Marques RB, Erkens-Schulze S, de Ridder CM, Hermans KG, Waltering K et al (2005) Androgen receptor modifications in prostate cancer cells upon long-termandrogen ablation and antiandrogen treatment. Int J Cancer 117:221–229
Marusyk A, Almendro V, Polyak K (2012) Intra-tumour heterogeneity: a looking glass for cancer? Nat Rev Cancer 12:323–334
Matei DV, Renne G, Pimentel M, Sandri MT, Zorzino L et al (2012) Neuroendocrine differentiation in castration-resistant prostate cancer: a systematic diagnostic attempt. Clin Genitourin Cancer 10(3):164–173
Mezynski J, Pezaro C, Bianchini D, Zivi A, Sandhu S et al (2012) Antitumour activity of docetaxel following treatment with the CYP17A1 inhibitor abiraterone: clinical evidence for cross-resistance? Ann Oncol 23(11):2943–2947
Miki J, Rhim JS (2008) Prostate cell cultures as in vitro models for the study of normal stem cells and cancer stem cells. Prostate Cancer Prostatic Dis 11:32–39
Milosevic M, Warde P, Menard C, Chung P, Toi A et al (2012) Tumor hypoxia predicts biochemical failure following radiotherapy for clinically localized prostate cancer. Clin Cancer Res 18:2108–2114
Mirchandani D, Zheng J, Miller GJ, Ghosh AK, Shibata DK et al (1995) Heterogeneity in intratumor distribution of p53 mutations in human prostate cancer. Am J Pathol 147:92–101
Montgomery RB, Mostaghel EA, Vessella R, Hess DL, Kalhorn TF et al (2008) Maintenance of intratumoral androgens in metastatic prostate cancer: a mechanism for castration-resistant tumor growth. Cancer Res 68:4447–4454
Moore N, Lyle S (2011) Quiescent, slow-cycling stem cell populations in cancer: a review of the evidence and discussion of significance. J Oncol, 11 pages. doi: 10.1155/2011/396076
Moore N, Houghton J, Lyle S (2012) Slow-cycling therapy-resistant cancer cells. Stem Cells Dev 21:1822–1830
Munoz J, Stange DE, Schepers AG, van de Wetering M, Koo BK et al (2012) The Lgr5 intestinal stem cell signature: robust expression of proposed quiescent '+4' cell markers. EMBO J 31(14):3079–3091
Navone NM, Labate ME, Troncoso P, Pisters LL, Conti CJ et al (1999) p53 mutations in prostate cancer bone metastases suggest that selected p53 mutants in the primary site define foci with metastatic potential. J Urol 161:304–308
Patrawala L, Calhoun T, Schneider-Broussard R, Li H, Bhatia B et al (2006) Highly purified CD44+ prostate cancer cells from xenograft human tumors are enriched in tumorigenic and metastatic progenitor cells. Oncogene 25:1696–1708
Patrawala L, Calhoun-Davis T, Schneider-Broussard R, Tang DG (2007) Hierarchical organization of prostate cancer cells in xenograft tumors: the CD44+alpha2beta1+ cell population is enriched in tumor-initiating cells. Cancer Res 67:6796–6805
Pece S, Tosoni D, Confalonieri S, Mazzarol G, Vecchi M et al (2010) Biological and molecular heterogeneity of breast cancers correlates with their cancer stem cell content. Cell 140:62–73
Phillips TM, McBride WH, Pajonk F (2006) The response of CD24(−/low)/CD44+ breast cancer-initiating cells to radiation. J Natl Cancer Inst 98:1777–1785
Piekarz RL, Bates SE (2009) Epigenetic modifiers: basic understanding and clinical development. Clin Cancer Res 15:3918–3926
Pietra G, Manzini C, Vitale M, Balsamo M, Ognio E et al (2009) Natural killer cells kill human melanoma cells with characteristics of cancer stem cells. Int Immunol 21:793–801

Pili R, Salumbides B, Zhao M, Altiok S, Qian D et al (2012) Phase I study of the histone deacetylase inhibitor entinostat in combination with 13-cis retinoic acid in patients with solid tumours. Br J Cancer 106:77–84
Qi Y, Li RM, Kong FM, Li H, Yu JP et al (2012) How do tumor stem cells actively escape from host immunosurveillance? Biochem Biophys Res Commun 420:699–703
Quintana E, Shackleton M, Sabel MS, Fullen DR, Johnson TM et al (2008) Efficient tumour formation by single human melanoma cells. Nature 456:593–598
Rane J, Pellacani D, Maitland NJ (2012) Advanced prostate cancer: a case for adjuvant differentiation therapy. Nat Rev Urol 9:595–602
Rehman Y, Rosenberg JE (2012) Abiraterone acetate: oral androgen biosynthesis inhibitor for treatment of castration-resistant prostate cancer. Drug Des Devel Ther 6:13–18
Reichert S, Rodel C, Mirsch J, Harter PN, Tomicic MT et al (2011) Survivin inhibition and DNA double-strand break repair: a molecular mechanism to overcome radioresistance in glioblastoma. Radiother Oncol 101:51–58
Reya T, Clevers H (2005) Wnt signalling in stem cells and cancer. Nature 434:843–850
Risbridger GP, Taylor RA (2008) Minireview: regulation of prostatic stem cells by stromal niche in health and disease. Endocrinology 149:4303–4306
Ruiz C, Lenkiewicz E, Evers L, Holley T, Robeson A et al (2011) Advancing a clinically relevant perspective of the clonal nature of cancer. Proc Natl Acad Sci USA 108:12054–12059
Salm SN, Burger PE, Coetzee S, Goto K, Moscatelli D et al (2005) TGF-{beta} maintains dormancy of prostatic stem cells in the proximal region of ducts. J Cell Biol 170:81–90
Sanchez P, Hernandez AM, Stecca B, Kahler AJ, DeGueme AM et al (2004) Inhibition of prostate cancer proliferation by interference with SONIC HEDGEHOG-GLI1 signaling. Proc Natl Acad Sci USA 101:12561–12566
Schroder F, Crawford ED, Axcrona K, Payne H, Keane TE (2012) Androgen deprivation therapy: past, present and future. BJU Int 109(Suppl 6):1–12
Scotto KW (2003) Transcriptional regulation of ABC drug transporters. Oncogene 22:7496–7511
Sebastian de Bono J, Sandhu S, Attard G (2011) Beyond hormone therapy for prostate cancer with PARP inhibitors. Cancer Cell 19:573–574
Seiwert TY, Salama JK, Vokes EE (2007) The concurrent chemoradiation paradigm–general principles. Nat Clin Pract Oncol 4:86–100
Sell S (2004) Stem cell origin of cancer and differentiation therapy. Crit Rev Oncol Hematol 51:1–28
Seruga B, Ocana A, Tannock IF (2011) Drug resistance in metastatic castration-resistant prostate cancer. Nat Rev Clin Oncol 8:12–23
Sharifi N, Kawasaki BT, Hurt EM, Farrar WL (2006) Stem cells in prostate cancer: resolving the castrate-resistant conundrum and implications for hormonal therapy. Cancer Biol Ther 5:901–906
Shi M, Guo XT, Shu MG, Chen FL, Li LW (2007) Cell-permeable hypoxia-inducible factor-1 (HIF-1) antagonists function as tumor radiosensitizers. Med Hypotheses 69:33–35
Siegel R, Desantis C, Virgo K, Stein K, Mariotto A et al (2012) Cancer treatment and survivorship statistics, 2012. CA Cancer J Clin 62(4):220–241
Skvortsova I, Skvortsov S, Stasyk T, Raju U, Popper BA et al (2008) Intracellular signaling pathways regulating radioresistance of human prostate carcinoma cells. Proteomics 8: 4521–4533
Snippert HJ, Clevers H (2011) Tracking adult stem cells. EMBO Rep 12:113–122
Sottocornola R, Lo Celso C (2012) Dormancy in the stem cell niche. Stem Cell Res Ther 3:10
Stark GR, Wahl GM (1984) Gene amplification. Annu Rev Biochem 53:447–491
Swift SL, Burns JE, Maitland NJ (2010) Altered expression of neurotensin receptors is associated with the differentiation state of prostate cancer. Cancer Res 70:347–356
Takebe N, Harris PJ, Warren RQ, Ivy SP (2011) Targeting cancer stem cells by inhibiting Wnt, Notch, and Hedgehog pathways. Nat Rev Clin Oncol 8:97–106

Tamada M, Nagano O, Tateyama S, Ohmura M, Yae T et al (2012) Modulation of glucose metabolism by CD44 contributes to antioxidant status and drug resistance in cancer cells. Cancer Res 72:1438–1448

Tanei T, Morimoto K, Shimazu K, Kim SJ, Tanji Y et al (2009) Association of breast cancer stem cells identified by aldehyde dehydrogenase 1 expression with resistance to sequential Paclitaxel and epirubicin-based chemotherapy for breast cancers. Clin Cancer Res 15:4234–4241

Trerotola M, Rathore S, Goel HL, Li J, Alberti S et al (2010) CD133, Trop-2 and alpha2beta1 integrin surface receptors as markers of putative human prostate cancer stem cells. Am J Transl Res 2:135–144

van der Pluijm G (2011) Epithelial plasticity, cancer stem cells and bone metastasis formation. Bone 48:37–43

Verheij M, Vens C, van Triest B (2010) Novel therapeutics in combination with radiotherapy to improve cancer treatment: rationale, mechanisms of action and clinical perspective. Drug Resist Updat 13:29–43

Visvader JE, Lindeman GJ (2008) Cancer stem cells in solid tumours: accumulating evidence and unresolved questions. Nat Rev Cancer 8:755–768

Vlashi E, Lagadec C, Vergnes L, Matsutani T, Masui K et al (2011) Metabolic state of glioma stem cells and nontumorigenic cells. Proc Natl Acad Sci USA 108:16062–16067

Waltering KK, Urbanucci A, Visakorpi T (2012) Androgen receptor (AR) aberrations in castration-resistant prostate cancer. Mol Cell Endocrinol 360(1–2):38–43

Wang Z, Li Y, Banerjee S, Kong D, Ahmad A et al (2010) Down-regulation of Notch-1 and Jagged-1 inhibits prostate cancer cell growth, migration and invasion, and induces apoptosis via inactivation of Akt, mTOR, and NF-kappaB signaling pathways. J Cell Biochem 109:726–736

Wicha MS (2008) Cancer stem cell heterogeneity in hereditary breast cancer. Breast Cancer Res 10:105

Wright NA (2012) Stem cell identification-in vivo lineage analysis versus in vitro isolation and clonal expansion. J Pathol 227:255–266

Xiao W, Graham PH, Power CA, Hao J, Kearsley JH et al (2012) CD44 is a biomarker associated with human prostate cancer radiation sensitivity. Clin Exp Metastasis 29:1–9

Yan M, Callahan CA, Beyer JC, Allamneni KP, Zhang G et al (2010) Chronic DLL4 blockade induces vascular neoplasms. Nature 463:E6–7

Zelivianski S, Verni M, Moore C, Kondrikov D, Taylor R et al (2001) Multipathways for transdifferentiation of human prostate cancer cells into neuroendocrine-like phenotype. Biochim Biophys Acta 1539:28–43

# Chapter 5
# Genetic and Signaling Pathway Regulations of Tumor-Initiating Cells of the Prostate

**David J. Mulholland and Hong Wu**

**Abstract** Here we review current literature on genetic and signaling pathway regulators of tumor initiating cells in prostate cancer. While we emphasize the consequence of PTEN loss and PI3K/AKT activation in prostate cancer initiating cells, we also assess the importance of other signaling regulators, including RAS/MAPK, WNT/β-catenin, MYC, NKX3.1 and p53 on these cells. Importantly, we stress how these factors alone, or in collaboration, alter tumor initiating cell/cancer stem cell function and consequently, phenotypes in in vivo prostate cancer models. Our review also highlights the understanding of how genetic pathway alteration influences cancer initiation by way of lineage tracing or cell type specific disruption. Functional similarities and differences of how tumor suppressor loss impacts human prostate cancer are also addressed, where appropriate. Finally, we touch on outstanding questions that future experimentation will hopefully address.

## 5.1 Introduction

A significant portion of men with prostate cancer will experience recurrent disease after successfully responding to front-line therapies (Grubb et al. 2007). Of those that receive androgen-deprivation therapy (ADT), many will progress to

D.J. Mulholland, M.S, Ph.D.
Department of Molecular and Medical Pharmacology, University of California, Los Angeles, CA 90095, USA
e-mail: dmulholland@mednet.ucla.edu

H. Wu, M.D., Ph.D. (✉)
Department of Molecular and Medical Pharmacology, University of California, Los Angeles, CA 90095, USA

Institute for Molecular Medicine, University of California, Los Angeles, CA 90095, USA

Eli and Edythe Broad Center of Regenerative Medicine and Stem Cell Research, University of California, Los Angeles, CA 90095, USA
e-mail: hwu@mednet.ucla.edu

S.D. Cramer (ed.), *Stem Cells and Prostate Cancer*, DOI 10.1007/978-1-4614-6498-3_5, 

castration-resistant prostate cancer (CRPC) (Kasper an. Cookson 2006). At clinical presentation, recurrent primary and metastatic prostate cancer is composed mostly of well-differentiated epithelium. This observation suggests the existence of unique populations of prostate cancer cells with the ability to survive therapy, remain quiescent for an extended period of time, and eventually differentiate to clinically detectable disease. Given this, the "holy grail" of prostate cancer treatment may depend on the identification of those genetic and pathway alterations responsible for the formation, or the maintenance, of these unique populations of cancer cells and ultimately leading to their elimination.

Though many terms have been used to describe these unique populations, for clarity we adapt the terminology defined by Dr. Visvader (Visvader and Lindeman 2008) and term those cells within the normal epithelium that acquire initial genetic or pathway alterations as tumor-initiating cells (TICs). Cells within the tumor mass that are capable of escaping therapies and responsible for repopulating tumor cells are referred to as cancer stem cells (CSCs) (Visvader and Lindeman 2008). Although a large body of published works based on prostate cancer cell lines, xenografts, in vitro spheroids cultures, genetically engineered mouse models (GEMs), and human cancer samples support the presence of prostate TICs and CSCs, few studies have resolved the relationship between TICs and CSCs in prostate cancer. This is partly due to a less well-defined lineage hierarchy of prostate stem cells in comparison to other somatic stem cells such as those of the hematopoietic system or mammary gland, as well as limited information on the major genetic and pathway alterations that are responsible for prostate cancer initiation, progression, and acquisition of therapeutic resistance. Recently developed model systems that allow lineage tracing and lineage-specific manipulation of key pathway alterations found in human prostate cancers have begun to provide clarity to the outline of the prostate stem cell hierarchy (Fig. 5.1).

Prostate cancer is genetically multifocal and frequently presents with aberrations in pathways that could regulate the activity of TICs/CSCs. This is exemplified by recent integrated genomic and mutational landscape studies, which have identified deletion/mutations of the PTEN, RB, p53 tumor suppressors, and heightened activities of PI3K/AKT, RAS/MAPK, RB/p53, AR, and WNT signaling pathways (Taylor et al. 2010; Barbieri et al. 2012; Grasso et al. 2012). Remarkably, PI3K/AKT and RAS pathway alterations are found in over 40% of primary tumors and over 90% in metastatic lesions while RB/p53 signaling is altered in more than 30% and 70% of primary and metastatic sites, respectively (Taylor et al. 2010). Studies have also implicated overexpression of the Polycomb members EZH2 and BMI-1 with late-stage prostate cancer (van Leenders et al. 2007; Ogata et al. 2004). Importantly, many of these mutations and their associated pathways are known to play essential roles in regulating TIC and CSC activities, including self-renewal and multi-lineage differentiation. These deregulated pathways may also represent potential targets for TIC and CSCs. Here we highlight evidence that such pathway alterations, elicited by certain genetic aberrations, can regulate the activity of TIC/CSC function during prostate cancer progression. We will summarize and interpret results from two experimental approaches, namely, in vitro manipulation of prostate cancer cell lines

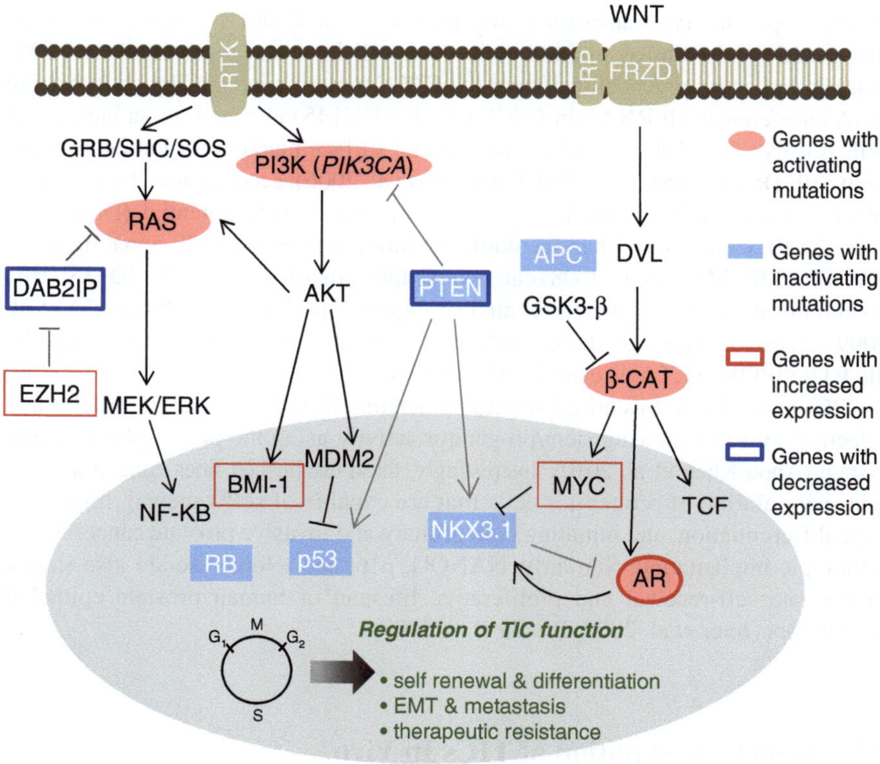


**Fig. 5.1** Genetic mutations and pathway alterations found in human prostate cancers that are known to regulate TIC function. Silencing of PTEN results in PI3K/AKT activation and NKX3.1, AR, and p53 suppression, leading to stem/progenitor cell expansion. RAS pathway activation occurs, in part, through increased EZH2 expression and reduced DAB2IP which may then act in concert with PI3K/AKT signaling to facilitate cancer progression, EMT, and metastasis. Cooperativity between loss p53 and RB can promote expansion of TICs through regulation of cell cycle checkpoints and potentially facilitate neuroendocrine differentiation. WNT/β-catenin pathway activation may regulate self-renewal through TCF activation but may also promote differentiation by coactivation of AR

and in vivo genetically engineered mouse models. Results from in vivo tissue recombination of murine and human origins will be reviewed by Drs. Goldstein and Witte (see Chap. 2).

### *5.1.1 Genetic Regulation of TIC/CSC In Vitro*

Increasing evidence indicates that many human prostate cancer cell lines and xenografts contain minor populations of cells, identifiable by either cell surface marker expression or enhanced drug efflux (Patrawala et al. 2005), with enhanced CSC function upon in vivo transplantation (Patrawala et al. 2006). Reports also suggest that genetic and pathway alterations in these cell lines can result in altered CSC number and activity. Knocking down the *PTEN* tumor suppressor by short hairpin RNA interference (sh-RNAi) in *PTEN*-positive DU145 cells leads to an increase in cells positive for $CD44^{+}$;$CD133^{+}$, putative cell surface markers for human prostate stem cells (Richardson et al. 2004; Patrawala et al. 2006), accompanied by enhanced sphere forming ability, clonal outgrowths and tumorigenic potential (Dubrovska et al. 2009). Conversely, pharmacological inhibition of PI3K/AKT signaling using NVPBEZ235 (Maira et al. 2008) can reverse the expansion of $CD44^{+}$;$CD133^{+}$ cells potentially mediated by increased nuclear expression of FOXO (Dubrovska et al. 2009). These data provide direct evidence that alteration of pathways regulated by the PTEN/PI3K/AKT axis can directly influence CSC activity.

TICs have also been studied in vitro by deleting floxed *Pte*n and *p53* alleles and selecting for self-renewing stem/progenitor activity using the prostosphere culture system (Abou-Kheir et al. 2010). Interestingly, these established lines contain a stable minor population of progenitor cells that are capable of self-renewal, multi-lineage differentiation, and initiating both primary and invasive prostate cancers upon orthotopic implantation. Similarly, NANOG, p16, and telomerase are also shown to regulate self-renewal and proliferative lifespan of human prostate epithelial progenitors (Jeter et al. 2009; Bhatia et al. 2008).

## 5.2 Genetic Regulation of TICs In Vivo

### *5.2.1 Phenotypes Associated with Genetic and Pathway Manipulations*

The development of genetically engineered mouse models (GEMs) that recapitulate major genetic or pathway alterations found in human prostate cancers has facilitated our understanding of the relationship of cell of origin, pathway alteration, and prostate cancer pathology. The first generation of transgenic models include TRAMP (Greenberg et al. 1995; Gingrich and Greenberg 1996) and LADY, which rely on the androgen-responsive probasin promoter to drive SV40 T-antigen expression, leading to the inhibition of both p53 and RB functions (Ahuja et al. 2005). The LADY model progresses to high-grade PIN while the TRAMP model progresses to invasive and metastatic disease accompanied by high content of neuroendocrine marker-positive cells. Using the same promoter, prostate-specific overexpression of the constitutively activated AKT (Majumder et al. 2003 PNAS 100:7841-6) and the MYC oncogene (Ellwood-Yen et al. 2003) leads to the development of PIN and invasive adenocarcinoma, respectively. They are potentially facilitated by the coinciding reduction of NKX3.1 expression (Iwata et al. 2010). However, no published studies have investigated the presence of either TICs or CSCs in these models. Moreover, it is unclear as to whether the different pathologies observed in these models are due to (1) a cell of origin, (2) specific genetic or pathway alterations, or (3) both.

The *Pten* conditional knockout prostate cancer model, driven by Cre4-Cre transgene (Wu et al. 2001), progresses to invasive adenocarcinoma and micrometastasis with well-defined kinetics (Wang et al. 2003) and has been used to study the effects of *Pten* deletion in both basal and luminal cells (Wang et al. 2006). *Pten* deletion leads to increased basal cell density, altered basal cell morphology, and localization (Wang et al. 2006), as supported by the expansion of $Cre^{+}$ and basal cell marker-positive cells along the basement membrane and migration of basal marker-positive cells to the lumen. *Pten* deletion also leads to increased cell proliferation and decreased cell death in the $CK5^{+}$;$CK8^{+}$;$P63^{-}$;$Ki67^{+}$;$AR^{+}$;BCL-$2^{+}$ population that is immediately adjacent to the basal compartment, which may be similar to the transit-amplifying (TA) cell population found in the human prostate (Wang et al. 2006; Vander Griend et al. 2008). Such observations are in parallel with the finding that PSCA, a marker associated with TA cells in human prostate cancer (Tran et al. 2002), is increased in *Pten*-deleted prostates (Dubey et al. 2001; Wang et al. 2003). Collectively, these data indicate that PTEN loss in the prostate epithelium can lead to the expansion of cells that function as TICs for prostate cancer development.

Greater understanding of the effects of pathway alterations on TIC has also been facilitated by the use of surface antigens to enrich for specific cell populations. $LSC^{high}$ cells (lineage marker negative [$CD31^{neg}CD45^{neg}Ter119^{neg}$] and Sca-$1^{+}$;$CD49f^{high}$) (Lawson et al. 2007), and more recently $LSC^{high}CD166^{high}$ cells (Jiao et al. 2012), are minor basal/TA cell populations within the wild-type prostate epithelium that harbor a high content of stem/progenitor activity and are enriched in the proximal region of the murine prostatic lobes (Fig. 5.2). In *Pten*-null prostates, Sca-$1^{+}$ cells are expanded by tenfold (Wang et al. 2006) and $LSC^{high}$ content increases by 2.5-fold during disease progression and further enhanced after castration (Mulholland et al. 2009; Liao et al. 2010). Interestingly, upregulation of the PI3K pathway and increased AR expression in human basal $LTC^{high}$ cells ($Lin^{-}$; $Trop2^{+}$;$CD49f^{high}$), which are the human equivalent of murine $LSC^{high}$ cells, are also necessary and sufficient for the initiation of human prostate cancer (Goldstein et al. 2010). Studies using in vivo transplants have demonstrated that the predominant tumor-regenerating cell population within the *Pten*-null model is also enriched in the $LSC^{high}$ but not $LSC^{low}$ subpopulations. This is true whether cells are prospectively isolated by flow cytometry (Goldstein et al. 2010; Lawson et al. 2010; Xin et al. 2005) and manipulated ex vivo or isolated directly from GEMs (Mulholland et al. 2009, 2012; Liao et al. 2010). $LSC^{high}$ cells also have the capacity to generate differentiated luminal cells upon transplantation (Mulholland et al. 2009, 2012; Goldstein et al. 2008, 2010).

At least four *Pten* conditional deletion lines (Backman et al. 2004; Wang et al. 2003; Chen et al. 2005; Ma et al. 2005) and two other prostate Cre lines (Ma et al. 2005; Maddison et al. 2000) have been generated by various laboratories. Of these models, not all conditional *Pten* prostate deletion models elicit the same phenotype. For instance, unlike the *Probasin-Cre4*$^{+}$;*Pten*$^{loxP/loxP}$ model (Wang et al. 2003), *PSA-Cre*$^{+}$;*Pten*$^{loxP/loxP}$ mutants (Ma et al. 2005) do not exhibit pathology in the proximal regions of the prostate but display accumulation of cells positive for the luminal markers CK8 and P-AKT. In this model, rare $Clu^{+}$;$Tacstd2^{+}$;Sca-$1^{+}$ cells may

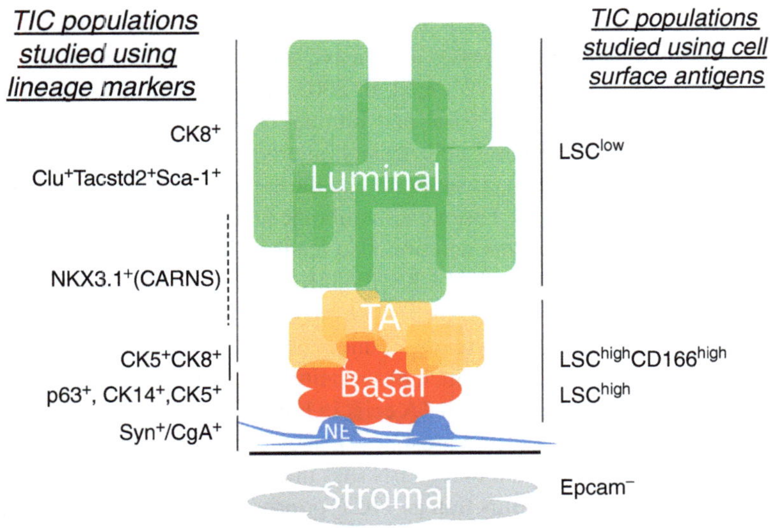


**Fig. 5.2** Lineage and cell surface markers used for identifying TICs in *Pten*-null murine prostate epithelium. The prostate epithelium consists of basal, transient-amplifying, luminal, and neuroendocrine cells. Basal cells are $p63^+$;$CK5^+$;$CK14^+$ while luminal cells are $CK8^+$. Intermediate or transient-amplifying (TA) cells include $CK5^+$;$CK8^+$ cells and may putatively include (*dashed line*) rare NKX3.1$^+$ (CARN) cells. Various cell surface markers have been used for isolating subpopulations with the epithelium, including $LSC^{high}$ ($Lin^-Sca\text{-}1^+CD49^{high}$), $LSC^{high}CD166^{high}$ ($Lin^-Sca\text{-}1^+CD49^{high}CD166^{high}$), and $LSC^{low}$ ($Lin^-Sca\text{-}1^+CD49^{low}$) subpopulations. Although these subpopulations correspond to the basal/TA and luminal cells in the WT epithelium, respectively, their localizations are altered in the *Pten*-null epithelium, especially after castration

constitute luminal TICs (Korsten et al. 2009) (Fig. 5.2). These data suggest that multiple stem/progenitor cells can serve as the cell of origin of prostate cancer initiated by PTEN loss and more defined systems, *via* lineage tracing and cell type-specific deletion, are necessary to characterize their role(s) in prostate cancer initiation and progression.

### 5.2.2 *Lineage Tracing for Identifying TICs*

Two general approaches have been used for identifying specific cell types associated to TICs, including lineage tracing and using cell type-specific promoters to drive pathway alterations. By crossing an *NKX3.1-Cre$^{ER}$* inducible line with the *Rosa26-floxed-LacZ* line, a rare population of cells called CARNs, or CAstration-Resistant Nkx3.1-expressing cells (Fig. 5.2), have been identified to have self-renewal capacity in vivo and can give rise to both basal and luminal cells. Targeted deletion of *Pten* driven by the *NKX3.1-Cre$^{ER}$* leads to the formation of invasive carcinoma and suggests that these bipotential CARN cells may be TICs (Wang et al. 2009). Using a similar lineage-tracing strategy but with more cell type-specific *Cre*$^{ER}$ lines, recent studies have provided more definitive proof that both basal and luminal prostate epithelial cells are self-sustainable lineages and can serve as a cell of origin for prostate cancer initiation and progression upon *Pten* deletion (Choi et al. 2012).

However, precautions need to be made when interpreting the current data concerning TICs in the prostate. Although CK5$^+$;CK8$^+$ double-positive TA cells represent a minor and almost undetectable population in the normal prostate epithelium, they may expand upon oncogenic insult, such as with *Pten* deletion (Wang et al. 2006). Interestingly, *Pten*-null lesions derived from the *CK5-Cre*$^{ER}$ promoter also yield CK5$^+$;CK8$^+$ double-positive TA cells and potentially to the expansion of CK8+ luminal cells. Since TA cells share both basal and luminal cell features, induced genetic or pathway alterations by cell type-specific Cre expression including, those driven by the *Nkx3.1*, *Ck5*, or *Ck8* promoters, may have differential effects on the currently poorly defined TA population. Second, the timing of genetic manipulation may influence TIC formation. While early deletion of *Pten* leads to micrometastatic disease, postpubescent deletion of *Pten* yields considerably more latent disease (Luchman et al. 2008), suggesting that the increased proliferation observed during normal prostate development may compound with oncogenic insults to affect cancer initiation, progression along with TIC content and activity. Nevertheless, the consensus is that multiple cell types or progenitor cells can serve as the cells of origin for prostate cancer development and become TICs upon PTEN loss. Future work should investigate whether such multi-TIC models apply to other genetic and pathway alterations and whether TIC cells from different cells of origin have similar or different capacity in cancer metastasis and CRPC development.

### 5.2.3 Cooperative Effects of Pathway Alteration on TIC Function

Multi-genetic events are known to collaboratively contribute to human cancer initiation, progression, metastasis, and therapeutic resistance. Recent studies suggest that the genetic events involved in "multiple-hit" tumorigenesis likely take place at the level of TICs (Guo et al. 2008). Similarly, second hits or multi-pathway alterations are frequently associated with enhanced progression in genetically defined prostate cancer GEMs. In some cases, this accelerated progression is associated with increased and sustained self-renewal activity of TICs (Martin et al. 2011). For example, advanced human prostate cancer frequently presents with PTEN and TP53 loss (Schlomm et al. 2008; Sircar et al. 2009). Mice with PTEN and TP53 loss (*Pb-Cre4*$^+$;*Pten*$^{loxP/loxP}$;*TP53*$^{loxP/loxP}$) recapitulate this collaborative effect with accelerated locally invasive disease (Chen et al. 2005; Abou-Kheir et al. 2010; Martin et al. 2011), which may be explained both by the reduced senescence observed in *Pten*-null *P53* null prostate cells (Chen et al. 2005) and the higher content of TICs observed in more aggressive cancers (Savona and Talpaz 2008; Pece et al. 2010). *Pten*-null;*TP53*-null prostate spheres and colonies displayed increased self-renewal capacity compared to spheres derived from *Pten*-null or WT prostates (Martin et al. 2011), observations that are consistent with findings that p53 loss leads to increased asymmetrical cell division in mammary stem cells (Cicalese et al. 2009). More recent studies also demonstrate the collaboration of JNK and the PI3K/AKT pathway in prostate cancer development. Downregulation of the JNK signaling pathway by way of conditional deletion of both *Jnk1/2* and upstream kinases *Mkk4/Mkk7* in *Pten*-null prostate epithelium leads to the expansion of $p63^+$ and $CD44^+$ cells and enhanced disease progression (Hubner et al. 2012).

Epithelial to mesenchymal transition, or EMT, is a phenotype associated both with increased cellular migration and stem cell function (Mani et al. 2008). Enhanced EMT-like qualities have been implicated in human prostate cancer recurrence (Zhang et al. 2009) and may also have potential as therapeutic targets (Tanaka et al. 2010). Activation of the RAS/MAPK signaling axis has been demonstrated to cooperate with PTEN loss in promoting EMT and macromatastasis in distant organs, accompanied by enhanced TIC function (Mulholland et al. 2012). In contrast to $LSC^{high}$ TIC of the *Pten*-null model (Mulholland et al. 2009), a new $EpCam^{low}$/$CD24^{low}$ EMT-like cells from mice with coordinate homozygous *Pten* loss and *K-ras* activation (*Pb-Cre4*$^+$;*Pten*$^{loxP/loxP}$;*K-ras*$^{G12D/WT}$) were revealed to have considerable TIC activity as compared to cells with only PTEN loss (Mulholland et al. 2012). Importantly, the ability of these EMT-like cells to initiate prostate carcinogenesis (Mulholland et al. 2012) and distant metastasis is correlated with their increased stem cell signature (Mulholland et al. 2012; Kong et al. 2010), suggesting that $EpCam^{low}$/$CD24^{low}$ EMT-like cells may represent a unique TIC population with metastatic potential, similar to what has been reported in breast (Mani et al. 2008) and pancreatic cancers (Hermann et al. 2007).

The correlation between cancer aggressiveness and TIC content has prompted the examination of how molecules known to enhance cancer progression may regulate stem cell functions. Besides its key function in antagonizing the PI3K/AKT pathway, PTEN is known for its function in negatively regulating stem cell self-renewal, proliferation, and survival (Groszer et al. 2001; Stiles et al. 2006). PTEN loss promotes $G_0$-G1 cell cycle transition of neural and hematopoietic stem cells (Groszer et al. 2006; Yilmaz et al. 2006; Zhang et al. 2006), which leads to expansion of stem cell pools. With increased expression during human prostate cancer progression, the Polycomb member EZH2 (Varambally et al. 2002) has been linked to the regulation of EMT (Cao et al. 2008) and stem cell function (Kamminga et al. 2006; Suva et al. 2009). Loss of the microRNA let-7 is not only inversely correlated with EZH2 expression in clinical prostate cancer specimens but can directly regulate EZH2 activity in culture with potential to inhibit clonal growths in prostate epithelial cell lines (Kong et al. 2012). BMI-1 is another Polycomb family member whose elevated expression in low-grade prostate cancer samples has been correlated with biochemical recurrence (van Leenders et al. 2007) and poor clinical outcome (Glinsky et al. 2005). BMI-1 overexpression leads to expansion of $p63^+$ cells in p53-null prostate sphere cells (Lukacs et al. 2010) while its knock-down led to impaired carcinogenesis driven by either FGF overexpression or *Pten* deletion in tissue regeneration assays (Lukacs et al. 2010). By extension, it is feasible that over expression of BMI-1 in the murine prostate could lead to invasive carcinoma when combined with *Pten* haploinsufficiency. In vitro observations show that AKT phosphorylates and activates BMI-1, which parallels the observations of increased BMI expression human prostate cancers (Nacerddine et al. 2012). These findings indicate that PTEN negatively regulates BMI-1 function (Fan et al. 2009) and provides an additional link between PTEN function and stem cell self-renewal.

Activation of WNT signaling has been associated with maintenance of stem cell function in many cancers (Clevers and Nusse 2012). Exogenous WNT3a can increase prostate sphere size and self-renewal activity both in the LNCaP and C4-2B human prostate cancer cell lines (Bisson and Prowse 2009) and in primary murine prostate epithelium while in turn be inhibited by Notch signaling (Shahi et al. 2011). While activating point mutations in β-catenin occur at a low frequency in human prostate cancer (Gerstein et al. 2002), murine GEMs with either activated β-catenin ($Pb\text{-}Cre^{+};Ctnnb^{\Delta ex3}$) or deletion of *Apc*, a component of the β-catenin destruction complex, results in localized expansion of $p63^{+}$ cells. When K-ras and β-catenin signaling are coactivated ($Pb\text{-}Cre^{+};Ctnnb^{\Delta ex3};K\text{-}rasV12$), the resulting mutant mice have increased $CD44^{+}$ cells and a paralleled increase in progression (Pearson et al. 2009). Collectively, these data suggest that Polycomb and WNT signaling maintain a critical balance in cells thought to be necessary for tumor initiation that occur upon genetic aberration.

Pathway-specific alterations occurring as a consequence of two genetic hits may also have an impact on lineage differentiation. For example, while *p53* or *Rb* loss alone is not sufficient in cancer development (Zhou et al. 2006), their coordinate deletion yields locally invasive disease as well as the expansion of neuroendocrine marker-positive cells in the proximal region deemed the stem cell niche of the murine prostate (Tsujimura et al. 2002).

## 5.3 Summary and Future Directions

The development of more sophisticated models, especially those from GEMs, has allowed control over pathway alterations in a cell type and temporally controlled manner. Such systems demonstrate the ability of basal, transit-amplifying, and luminal cells to serve as TIC populations. These exciting findings now pave the way for investigators to dissect which cell populations may belong to the "holy grail" of cancer progression and to determine whether such cells may be targetable with currently available therapeutics.

Among many urgent issues need to be addressed, we suggest three:

1. To study how complimentary genetic and pathway alterations collaborate in regulating TICs and CSCs. Integrated genomic and mutation landscape studies will provide rich genetic and pathway information related to prostate cancer development, as exemplified by the presence of TMPRSS2-ERG fusion events coinciding with PTEN loss in human prostate cancer (Squire 2009) and their

cooperability in promoting progression (King et al. 2009). Future studies should consider whether such interactions may take place in stem/progenitor cells and in TIC formation. Additional molecular subtypes of human prostate, such as those characterized by *CHD1*, *SPOP*, *FOXA1*, and *MED12* mutations (Barbieri et al. 2012), should also be evaluated for their functional impact on TIC activity and disease progression.

2. To investigate both cell-intrinsic and cell-extrinsic mechanisms of TIC regulation. Studies of normal stem cells indicate that stem cell properties, including self-renewal and multi-lineage differentiation and homeostasis, are influenced by their local microenvironment or "niche." Alterations of the microenvironment can drive normal stem cells to become TICs or CSCs. Although most of the current studies focus on the cell-intrinsic regulation of TICs and CSCs, future studies should consider TIC and CSC activities with a "system" approach, especially in an immune-competent setting and within tumor's native environment or niche.
3. To determine whether, subsequent to initiation, CSCs are functionally important for progression, metastasis, and therapeutic resistance. Despite the multifocal and heterogeneous nature of human primary prostate cancers, where distinct molecular and genetic alterations are associated with various "clones," most metastatic prostate cancers and CRPC is monoclonal in origin, suggesting that metastatic lesions in different anatomical sites of the same patient may arise from a single precursor cell within the primary lesions (Grasso et al. 2012; Liu et al. 2009; Holcomb et al. 2009). Determing which genetic and pathway alterations lead to the formation of CSCs within the primary tumors and are responsible for metastasis and CRPC will be critical for the understanding and treatment of lethal prostate cancers.

## References


Abou-Kheir WG et al (2010) Characterizing the contribution of stem/progenitor cells to tumorigenesis in the Pten−/−TP53−/− prostate cancer model. Stem Cells 28(12):2129–2140

Ahuja D, Saenz-Robles MT, Pipas JM (2005) SV40 large T antigen targets multiple cellular pathways to elicit cellular transformation. Oncogene 24(52):7729–7745

Backman SA et al (2004) Early onset of neoplasia in the prostate and skin of mice with tissue-specific deletion of Pten. Proc Natl Acad Sci USA 101(6):1725–1730

Barbieri CE et al (2012) Exome sequencing identifies recurrent SPOP, FOXA1 and MED12 mutations in prostate cancer. Nat Genet 44(6):685–689

Bhatia B et al (2008) Critical and distinct roles of p16 and telomerase in regulating the proliferative life span of normal human prostate epithelial progenitor cells. J Biol Chem 283(41):27957–27972

Bisson I, Prowse DM (2009) WNT signaling regulates self-renewal and differentiation of prostate cancer cells with stem cell characteristics. Cell Res 19(6):683–697

Cao Q et al (2008) Repression of E-cadherin by the polycomb group protein EZH2 in cancer. Oncogene 27(58):7274–7284

Chen Z et al (2005) Crucial role of p53-dependent cellular senescence in suppression of Pten-deficient tumorigenesis. Nature 436(7051):725–730

Choi N et al (2012) Adult murine prostate basal and luminal cells are self-sustained lineages that can both serve as targets for prostate cancer initiation. Cancer Cell 21(2):253–265
Cicalese A et al (2009) The tumor suppressor p53 regulates polarity of self-renewing divisions in mammary stem cells. Cell 138(6):1083–1095
Clevers H, Nusse R (2012) Wnt/beta-catenin signaling and disease. Cell 149(6):1192–1205
Collins AT et al (2005) Prospective identification of tumorigenic prostate cancer stem cells. Cancer Res 65(23):10946–10951
Dubey P et al (2001) Alternative pathways to prostate carcinoma activate prostate stem cell antigen expression. Cancer Res 61(8):3256–3261
Dubrovska A et al (2009) The role of PTEN/Akt/PI3K signaling in the maintenance and viability of prostate cancer stem-like cell populations. Proc Natl Acad Sci USA 106(1):268–273
Ellwood-Yen K et al (2003) Myc-driven murine prostate cancer shares molecular features with human prostate tumors. Cancer Cell 4(3):223–238
Fan C et al (2009) PTEN inhibits BMI1 function independently of its phosphatase activity. Mol Cancer 8:98
Gerstein AV et al (2002) APC/CTNNB1 (beta-catenin) pathway alterations in human prostate cancers. Genes Chromosomes Cancer 34(1):9–16
Gingrich JR, Greenberg NM (1996) A transgenic mouse prostate cancer model. Toxicol Pathol 24(4):502–504
Glinsky GV, Berezovska O, Glinskii AB (2005) Microarray analysis identifies a death-from-cancer signature predicting therapy failure in patients with multiple types of cancer. J Clin Invest 115(6):1503–1521
Goldstein AS et al (2008) Trop2 identifies a subpopulation of murine and human prostate basal cells with stem cell characteristics. Proc Natl Acad Sci USA 105(52):20882–20887
Goldstein AS et al (2010) Identification of a cell of origin for human prostate cancer. Science 329(5991):568–571
Goto K et al (2006) Proximal prostatic stem cells are programmed to regenerate a proximal-distal ductal axis. Stem Cells 24(8):1859–1868
Grasso CS et al (2012) The mutational landscape of lethal castration-resistant prostate cancer. Nature 487(7406):239–243
Greenberg NM et al (1995) Prostate cancer in a transgenic mouse. Proc Natl Acad Sci USA 92(8):3439–3443
Groszer M et al (2001) Negative regulation of neural stem/progenitor cell proliferation by the Pten tumor suppressor gene in vivo. Science 294(5549):2186–2189
Groszer M et al (2006) PTEN negatively regulates neural stem cell self-renewal by modulating G0-G1 cell cycle entry. Proc Natl Acad Sci USA 103(1):111–116
Grubb RL 3rd, Kibel AS (2007) Prostate cancer: screening, diagnosis and management in 2007. Mo Med 104(5):408–413; quiz 413–414
Guo W et al (2008) Multi-genetic events collaboratively contribute to Pten-null leukaemia stem-cell formation. Nature 453(7194):529–533
Hermann PC et al (2007) Distinct populations of cancer stem cells determine tumor growth and metastatic activity in human pancreatic cancer. Cell Stem Cell 1(3):313–323
Holcomb IN et al (2009) Comparative analyses of chromosome alterations in soft-tissue metastases within and across patients with castration-resistant prostate cancer. Cancer Res 69(19): 7793–7802
Hubner A et al (2012) JNK and PTEN cooperatively control the development of invasive adenocarcinoma of the prostate. Proc Natl Acad Sci USA 109(30):12046–12051
Iwata T et al (2010) MYC overexpression induces prostatic intraepithelial neoplasia and loss of Nkx3.1 in mouse luminal epithelial cells. PLoS One 5(2):e9427
Jeter CR et al (2009) Functional evidence that the self-renewal gene NANOG regulates human tumor development. Stem Cells 27(5):993–1005
Jiao J et al (2012) Identification of CD166 as a Surface Marker for Enriching Prostate Stem/Progenitor and Cancer Initiating Cells. PLoS One 7(8):e42564

Kamminga LM et al (2006) The Polycomb group gene Ezh2 prevents hematopoietic stem cell exhaustion. Blood 107(5):2170–2179
Kasper S, Cookson MS (2006) Mechanisms leading to the development of hormone-resistant prostate cancer. Urol Clin North Am 33(2):201–210, vii
King JC et al (2009) Cooperativity of TMPRSS2-ERG with PI3-kinase pathway activation in prostate oncogenesis. Nat Genet 41(5):524–526
Kong D et al (2010) Epithelial to mesenchymal transition is mechanistically linked with stem cell signatures in prostate cancer cells. PLoS One 5(8):e12445
Kong D et al (2012) Loss of let-7 up-regulates EZH2 in prostate cancer consistent with the acquisition of cancer stem cell signatures that are attenuated by BR-DIM. PLoS One 7(3):e33729
Korsten H et al (2009) Accumulating progenitor cells in the luminal epithelial cell layer are candidate tumor initiating cells in a Pten knockout mouse prostate cancer model. PLoS One 4(5):e5662
Lawson DA et al (2007) Isolation and functional characterization of murine prostate stem cells. Proc Natl Acad Sci USA 104(1):181–186
Lawson DA et al (2010) Basal epithelial stem cells are efficient targets for prostate cancer initiation. Proc Natl Acad Sci USA 107(6):2610–2615
Liao CP et al (2010) Cancer-associated fibroblasts enhance the gland-forming capability of prostate cancer stem cells. Cancer Res 70(18):7294–7303
Liu W et al (2009) Copy number analysis indicates monoclonal origin of lethal metastatic prostate cancer. Nat Med 15(5):559–565
Luchman HA et al (2008) The pace of prostatic intraepithelial neoplasia development is determined by the timing of Pten tumor suppressor gene excision. PLoS One 3(12):e3940
Lukacs RU et al (2010) Bmi-1 is a crucial regulator of prostate stem cell self-renewal and malignant transformation. Cell Stem Cell 7(6):682–693
Ma X et al (2005) Targeted biallelic inactivation of Pten in the mouse prostate leads to prostate cancer accompanied by increased epithelial cell proliferation but not by reduced apoptosis. Cancer Res 65(13):5730–5739
Maddison LA et al (2000) Prostate specific expression of Cre recombinase in transgenic mice. Genesis 26(2):154–156
Maira SM et al (2008) Identification and characterization of NVP-BEZ235, a new orally available dual phosphatidylinositol 3-kinase/mammalian target of rapamycin inhibitor with potent in vivo antitumor activity. Mol Cancer Ther 7(7):1851–1863
Mani SA et al (2008) The epithelial-mesenchymal transition generates cells with properties of stem cells. Cell 133(4):704–715
Martin P et al (2011) Prostate epithelial Pten/TP53 loss leads to transformation of multipotential progenitors and epithelial to mesenchymal transition. Am J Pathol 179(1):422–435
Mulholland DJ et al (2009) Lin-Sca-1+ CD49fhigh stem/progenitors are tumor-initiating cells in the Pten-null prostate cancer model. Cancer Res 69(22):8555–8562
Mulholland DJ et al (2012) Pten loss and RAS/MAPK activation cooperate to promote EMT and metastasis initiated from prostate cancer stem/progenitor cells. Cancer Res 72(7):1878–1889
Nacerddine K et al (2012) Akt-mediated phosphorylation of Bmi1 modulates its oncogenic potential, E3 ligase activity, and DNA damage repair activity in mouse prostate cancer. J Clin Invest 122(5):1920–1932
Ogata R et al (2004) Identification of polycomb group protein enhancer of zeste homolog 2 (EZH2)-derived peptides immunogenic in HLA-A24+ prostate cancer patients. Prostate 60(4):273–281
Patrawala L et al (2005) Side population is enriched in tumorigenic, stem-like cancer cells, whereas ABCG2+ and ABCG2- cancer cells are similarly tumorigenic. Cancer Res 65(14):6207–6219
Patrawala L et al (2006) Highly purified CD44+ prostate cancer cells from xenograft human tumors are enriched in tumorigenic and metastatic progenitor cells. Oncogene 25(12):1696–1708
Pearson HB, Phesse TJ, Clarke AR (2009) K-ras and Wnt signaling synergize to accelerate prostate tumorigenesis in the mouse. Cancer Res 69(1):94–101
Pece S et al (2010) Biological and molecular heterogeneity of breast cancers correlates with their cancer stem cell content. Cell 140(1):62–73

Richardson GD et al (2004) CD133, a novel marker for human prostatic epithelial stem cells. J Cell Sci 117(Pt 16):3539–3545
Savona M, Talpaz M (2008) Getting to the stem of chronic myeloid leukaemia. Nat Rev Cancer 8(5):341–350
Schlomm T et al (2008) Clinical significance of p53 alterations in surgically treated prostate cancers. Mod Pathol 21(11):1371–1378
Shahi P et al (2011) Wnt and Notch pathways have interrelated opposing roles on prostate progenitor cell proliferation and differentiation. Stem Cells 29(4):678–688
Sircar K et al (2009) PTEN genomic deletion is associated with p-Akt and AR signalling in poorer outcome, hormone refractory prostate cancer. J Pathol 218(4):505–513
Squire JA (2009) TMPRSS2-ERG and PTEN loss in prostate cancer. Nat Genet 41(5):509–510
Stiles BL et al (2006) Selective deletion of Pten in pancreatic beta cells leads to increased islet mass and resistance to STZ-induced diabetes. Mol Cell Biol 26(7):2772–2781
Suva ML et al (2009) EZH2 is essential for glioblastoma cancer stem cell maintenance. Cancer Res 69(24):9211–9218
Tanaka H et al (2010) Monoclonal antibody targeting of N-cadherin inhibits prostate cancer growth, metastasis and castration resistance. Nat Med 16(12):1414–1420
Taylor BS et al (2010) Integrative genomic profiling of human prostate cancer. Cancer Cell 18(1):11–22
Tran CP et al (2002) Prostate stem cell antigen is a marker of late intermediate prostate epithelial cells. Mol Cancer Res 1(2):113–121
Tsujimura A et al (2002) Proximal location of mouse prostate epithelial stem cells: a model of prostatic homeostasis. J Cell Biol 157(7):1257–1265
van Leenders GJ et al (2007) Polycomb-group oncogenes EZH2, BMI1, and RING1 are overexpressed in prostate cancer with adverse pathologic and clinical features. Eur Urol 52(2):455–463
Vander Griend DJ et al (2008) The role of CD133 in normal human prostate stem cells and malignant cancer-initiating cells. Cancer Res 68(23):9703–9711
Varambally S et al (2002) The polycomb group protein EZH2 is involved in progression of prostate cancer. Nature 419(6907):624–629
Visvader JE, Lindeman GJ (2008) Cancer stem cells in solid tumours: accumulating evidence and unresolved questions. Nat Rev Cancer 8(10):755–768
Wang JC (2007) Evaluating therapeutic efficacy against cancer stem cells: new challenges posed by a new paradigm. Cell Stem Cell 1(5):497–501
Wang S et al (2003) Prostate-specific deletion of the murine Pten tumor suppressor gene leads to metastatic prostate cancer. Cancer Cell 4(3):209–221
Wang S et al (2006) Pten deletion leads to the expansion of a prostatic stem/progenitor cell subpopulation and tumor initiation. Proc Natl Acad Sci USA 103(5):1480–1485
Wang X et al (2009) A luminal epithelial stem cell that is a cell of origin for prostate cancer. Nature 461(7263):495–500
Wu X et al (2001) Generation of a prostate epithelial cell-specific Cre transgenic mouse model for tissue-specific gene ablation. Mech Dev 101(1–2):61–69
Xin L, Lawson DA, Witte ON (2005) The Sca-1 cell surface marker enriches for a prostate-regenerating cell subpopulation that can initiate prostate tumorigenesis. Proc Natl Acad Sci USA 102(19):6942–6947
Yilmaz OH et al (2006) Pten dependence distinguishes haematopoietic stem cells from leukaemia-initiating cells. Nature 441(7092):475–482
Zhang J et al (2006) PTEN maintains haematopoietic stem cells and acts in lineage choice and leukaemia prevention. Nature 441(7092):518–522
Zhang Q et al (2009) Nuclear factor-kappaB-mediated transforming growth factor-beta-induced expression of vimentin is an independent predictor of biochemical recurrence after radical prostatectomy. Clin Cancer Res 15(10):3557–3567
Zhou Z et al (2006) Synergy of p53 and Rb deficiency in a conditional mouse model for metastatic prostate cancer. Cancer Res 66(16):7889–7898

# Chapter 6
# The Prostate Stem Cell Niche

**David Moscatelli and E. Lynette Wilson**

**Abstract** Stem cells reside in a localized region called a niche where interactions with surrounding cells, especially stromal cells, maintain stem cell quiescence and multipotency and guide their transition to proliferation and differentiation. The mouse prostate stem cell niche has been localized to the proximal region of the gland, while the human niche is more diffuse. Stromal cells that may contribute signals to the niche include smooth muscle cells, fibroblasts, vascular endothelial cells, Schwann cells, and adipocytes. Signaling molecules that the stromal cells and stem cells use to communicate with each other include members of the BMP/TGF-ß, Wnt/ß-catenin, FGF, notch, hedgehog, and ephrin signaling pathways. Imbalances in signaling between the stroma and stem cells in the niche can lead to excessive proliferation of both epithelial and stromal components. Knowledge of the specific signals used and their source and targets may help decipher the complex interactions that keep prostate stem cells functioning normally and result in targeted therapies for treating benign and malignant prostatic proliferation.

D. Moscatelli, Ph.D. (✉)
Department of Cell Biology, and New York University Cancer Institute,
New York University School of Medicine, New York, NY, USA
e-mail: david.moscatelli@med.nyu.edu

E.L. Wilson, Ph.D.
Department of Cell Biology, and New York University Cancer Institute,
New York University School of Medicine, New York, NY, USA

Department of Urology, New York University School of Medicine,
New York, NY, USA

Helen L. and Martin S. Kimmel Center for Stem Cell Biology, New York University
School of Medicine, New York, NY, USA

S.D. Cramer (ed.), *Stem Cells and Prostate Cancer*, DOI 10.1007/978-1-4614-6498-3_6, 

## 6.1 Introduction

Organ-specific stem cells are cells that are able to generate all of the differentiated cell lineages of that organ. Under normal conditions, these cells are deeply quiescent, rarely replicating. In response to a loss of tissue, homeostatic signals activate a subset of these cells to divide. Cell division will result in two alternate fates: some daughter cells give rise to new stem cells to replenish the stem cell pool, while others differentiate into transit amplifying cells that proliferate further and differentiate to regenerate the lost cells. Once activated, stem cells are highly proliferative. Thus, their proliferation must be tightly regulated to prevent overgrowth of the tissue and to prevent exhaustion of the stem cell pool. A niche is a localized region of an organ that nurtures stem cells keeping them in an undifferentiated state and regulating their differentiation into more mature cells. It is the role of the niche to maintain stem cell quiescence and regulate their differentiation.

The concept of a niche arose from studies of hematopoietic stem cells. The niche for hematopoietic stem cells is located in the bone marrow and requires interactions between the stem cells and osteoblasts and/or vascular endothelial cells for the proper balance between stem cell renewal and differentiation (Orford and Scadden 2008). Localized niches have also been identified in the bulge region of the hair follicle, the base of intestinal glands, and the neck of stomach glands. Now restricted niches have been identified for a variety of tissue-specific stem cells, including corneal, skin, intestinal, neuronal, and prostate stem cells (Lavker et al. 2004; Fuchs and Horsley 2008; Clevers 2009; Miller and Gauthier-Fisher 2009; Tsujimura et al. 2002).

Stem cell niches keep cells with high proliferative potential quiescent. In tumors, cells with high proliferative potential proliferate without constraint. As tumor-initiating cells are thought to have properties in common with their organ-specific stem cell counterparts, knowledge of the mechanisms by which the niche controls stem cell renewal and differentiation may provide insights into targets that may be utilized to control tumor proliferation. Therefore, there has been much interest in defining the stem cell niche and the molecular signals between the niche cells and stem cells that regulate stem cell activity.

## 6.2 Evidence for a Prostatic Niche

The first evidence for a niche that supports stem cell function in the prostate came from the transplant experiments of Cunha and his coworkers. They observed that when the tips of prostate ducts were transplanted under the renal capsule of recipient mice, growth of prostate-like tissue occurred only when urogenital sinus mesenchyme (UGM) was co-implanted (Norman et al. 1986). Indeed, the amount of growth of the epithelial tissue depended on the amount of UGM added (Chung and Cunha 1983; Goto et al. 2006). Implantation of the tips alone or co-implantation of mesenchyme from the bladder in place of the UGM did not result in growth of prostate-like tissues (Norman et al. 1986). These experiments suggested that important signals for prostate development came from the stroma and that prostate stroma alone contained the signals to guide prostate development. Co-implantation of human prostate epithelial cells with rat UGM under the renal capsule also resulted in formation of prostate duct-like structures, suggesting that human and mouse stem cells respond to similar signals from the stroma (Hayward et al. 1998).

Further experiments showed that UGM was not only supportive of prostate development but was also instructive. When bladder epithelium was co-implanted with UGM under the renal capsule, prostate-like growths were produced. These growths synthesized prostatic proteins, demonstrating that full transdifferentiation into prostatic epithelium had occurred (Cunha et al. 1980, 1983; Neubauer et al. 1983). Similarly, when vaginal epithelium from female embryos was co-implanted with male UGM, the female epithelium formed prostate-like structures (Boutin et al. 1991). Thus, signals from the stroma directed undifferentiated epithelial cells in the bladder or female urogenital sinus to differentiate into functional prostatic epithelium.

## 6.3 Localization of the Adult Niche

In the mouse, two different niches can be defined. The first is the embryonic niche, in the urogenital sinus, which consists of the UGM and undifferentiated urogenital sinus epithelial cells that signal to each other to guide differentiation into adult prostatic tissue containing the proper balance of basal and luminal epithelial and neuroendocrine cells. The UGM is quite potent in guiding prostate development and can even direct the development of adult prostate stem cells (Xin et al. 2003; Goto et al. 2006). There have been no attempts to determine if the stem cell-supporting activity of the UGM can be further localized, but given the small size of the tissue, it is likely that the entire UGM has stem cell-supporting activity. The second niche is the adult stem cell niche. This niche is responsible for maintaining multipotent adult stem cells and regulating their differentiation in response to fluctuations in androgen levels.

Although the unfractionated UGM could act as a niche during development and in transplant experiments, the stem cell niche in adult prostate was expected to be more restricted. To map the adult niche in the mouse, advantage was taken of one characteristic of stem cells, their slow turnover. In this approach, the cells of prostates of young mice were uniformly labeled with bromodeoxyuridine, and the mice were subjected to multiple rounds of prostate involution and regeneration (Tsujimura et al. 2002). The cells that retained the bromodeoxyuridine label after this cycling of the prostates were presumed to be slowly turning over stem cells. These label-retaining cells were found predominately in the proximal region of the prostatic ducts. Later experiments found that prostate cells expressing high levels the cell-surface antigen Sca-1 (Sca-1$^{high}$) had two other properties of stem cells: a high growth potential and the ability to regenerate prostate-like organs when implanted under the renal capsule of recipient mice. The Sca-$1^{high}$ cells with high growth and regenerative capacity were preponderantly located in the proximal region (Burger et al. 2005; Xin et al. 2005). Interestingly, the proximal region of the regenerated prostate-like organs also contains cells with high growth capacity and greater ability to regenerate prostate-like organs when implanted under the renal capsule, suggesting that the stem cells are able to regenerate their niche (Goto et al. 2006).

Thus, the proximal region of the prostate gland contains cells with three attributes of stem cells, slow cycling, high growth potential, and prostate regenerating ability. Recent studies have suggested that stem cells may reside in both the basal cell and luminal cell compartments of the prostate epithelium (Wang et al. 2009; Goldstein et al. 2010; Lawson et al. 2010). As the slow cycling cells were localized in both the luminal and basal compartments of the proximal region with approximately equal frequency (Tsujimura et al. 2002), the proximal region may be the niche for both types of stem cells.

In humans, the niche appears not to be localized. Cells expressing CD133 and high levels of $\alpha_2\beta_1$ integrin have high proliferative potential and can regenerate prostate duct-like structures when implanted into immunocompromized mice, two attributes of stem cells. These rare $\alpha_2\beta_1$ integrin$^{high}$/CD133$^+$ cells are found scattered throughout the prostatic ducts often at the base of budding or branching regions (Richardson et al. 2004). While the human niche is not restricted in a particular region of the organ, it seems to be limited to highly characteristic microdomains of the organ.

## 6.4 Cells That Contribute to the Niche

The characteristics of the proximal region that specify it as the prostate stem cell niche must consist of the peculiar mix of signals from stromal cells found there (Fig. 6.1). There have been several attempts to identify the stromal cells that contribute to the support of stem cells. As the proximal ducts have a thicker investment with smooth muscle cells than other regions of the ducts (Nemeth and Lee 1996), smooth muscle cells may make an important contribution to the niche. In experiments co-implanting smooth muscle cells with prostate epithelial cells under the renal capsule, the smooth muscle cells supported the growth of the epithelial cells and their organization into duct-like structures, whereas no growth was observed in the absence of smooth muscle cells (Takao et al. 2003). In addition, when prostate epithelial cells were implanted with UGM cells the epithelial cells induced abundant differentiation of smooth muscle from the UGM, suggesting reciprocal signals between these compartments (Cunha et al. 1996). Although the smooth muscle cells can support limited duct formation, they are not able to support the formation of full prostate-like organs as UGM does.

The stem cells in many organs are closely associated with the vasculature. Slow cycling, label-retaining cells in the endometrium were found in close association

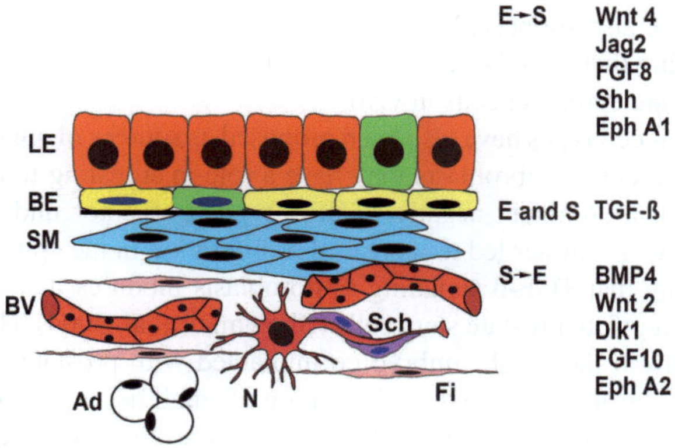


**Fig. 6.1** Epithelial–stromal interactions in the prostate stem cell niche. In the stem cell niche, luminal (*LE*, *orange*) and basal (*BE*, *yellow*) epithelial cells send signals that support the maintenance of a specialized stroma. Stem cells (*green*) that have been identified in both epithelial layers contribute to these signals. In turn, stromal cells, including smooth muscle cells (*SM*, *blue*), fibroblasts (*Fi*, *pink*), cells of the blood vessels (*BV*, *red*), neurons (*N*, *magenta*) with their associated Schwann cells (*Sch*, *purple*), and adipocytes (*Ad*, *white*), send signals that regulate the balance between quiescence, proliferation, and differentiation in the epithelial stem cells. Some epithelial-derived molecules that signal to the stroma or stroma-derived molecules that signal to the epithelium are listed on the right. Some signaling molecules, such as TGF-ß, are produced by both epithelium and stroma and act on both layers

with blood vessels (Chan and Gargett 2006), and cells with molecular markers of mesenchymal stem cells were identified in vascular walls (Shi and Gronthos 2003). In the adult hippocampus, proliferating neuronal progenitor cells were found to be preferentially associated with remodeling vasculature with proliferating endothelial cells (Palmer et al. 2000; Ohab et al. 2006). Similarly, cells with the molecular markers of hematopoietic stem cells have been observed in close association with vascular sinuses in the bone marrow and spleen (Yin and Li 2006; Kiel et al. 2005). In addition, knockout of stem cell factor expression in vascular endothelium inhibited hematopoiesis, suggesting that the endothelial cells are an important source of signals to stem cells in the bone marrow hematopoietic niche (Ding et al. 2012). Finally, in the testes, spermatogonia were also found associated with blood vessels, and alteration of the course of the vessels caused a redistribution of the spermatogonia so that they realigned with the new position of the vessels (Yoshida et al. 2007). These observations suggest that vascular endothelial cells impart signals that are important for the organization and function of stem cells. The proximal region of the prostate is also highly vascularized with twice the vessel density of the intermediate region and eight times the density of the distal region (Wang et al. 2007b). Prostatic epithelial growth and regression are tightly correlated with the growth or regression of the prostatic vasculature (Folkman 1998; Franck-Lissbrant et al. 1998; Shabisgh et al. 1999; English et al. 1985; Lissbrant et al. 2004; Wang et al. 2007a), suggesting that that there may be important signals between prostate stem cells and the vasculature. Further, co-implanting vascular endothelial cells with prostate epithelial cells under the renal capsule also supported the formation of duct-like structures (Bates et al. 2008), suggesting that endothelial cells also contribute signals that can support the growth of prostate epithelial cells in vivo.

Other stromal cell types have also been proposed as sources of signals to expanding prostate epithelium. Fibroblasts may have a role in signaling to the stem cells, as expression of a dominant-negative TGF-ß type II receptor under control of a fibroblast-specific promoter led to increased proliferation in the epithelial compartment, suggesting that TGF-ß signaling to fibroblasts modifies their production of molecules that regulate prostate stem cells (Bhowmick et al. 2004). However, fibroblasts derived from the whole embryo co-implanted with prostate epithelial cells under the renal capsule supported the formation of small tissue growths but did not support duct formation (Hayward et al. 1999). This result may indicate that fibroblasts alone do not produce an appropriate range of regulatory molecules to support prostate differentiation or a more specific subset of fibroblasts from the proximal prostate are needed. In addition, even adipocytes have been reported to support the growth of prostatic tissue in vitro, suggesting that these cells may also contribute to prostate stem cell proliferation and differentiation (Tokuda et al. 1999). Finally, as nonmyelinating Schwann cells wrapping sympathetic neurons help regulate the stem cell niche of hematopoietic cells (Yamazaki et al. 2011; Kunisaki and Frenette 2012), they may play a similar role in the prostate niche. The relative importance of signals from smooth muscle cells, vascular endothelial cells, fibroblasts, adipocytes, and Schwann cells in prostate stem cell function (Fig. 6.1) is difficult to ascertain, as these cells have never been directly compared in the same assay. Moreover, none of these cell types seems to be as efficient as UGM in supporting stem cell growth in vivo, suggesting that the niche requires a concert of signals contributed by multiple cell types, and fractionating the cells may disrupt important interactions that are necessary for a functioning niche. Perhaps a more sophisticated analysis is needed in which specific signaling molecules are knocked out or overexpressed in each of the stromal cell types in order to judge the contribution of a specific cell type to stem cell quiescence, proliferation, and differentiation.

## 6.5 Niche Signaling Molecules

There is much interest in the molecules used by stem cells and the specific stroma in their niche to signal to each other. These signaling molecules may provide a point of entry to the manipulation of stem cell quiescence and proliferation. Initial experiments have focused on the murine stem cell niche. However, as experiments with more purified human prostatic stem cells have confirmed that rat UGM will support the growth of human cells in transplant experiments (Goldstein et al. 2008), the signals used in rodents are likely to be applicable to manipulation of the human stem cell niche.

To obtain information on the signaling molecules used in the stem cell niche, molecules expressed by the urogenital sinus of the embryonic developing prostate were examined. Urogenital sinus epithelial cells (UGE), containing primitive prostate stem cells, and UGM cells were isolated from 16-day-old embryos, just prior to the invasion of the primitive prostatic buds into the surrounding mesenchyme (Blum et al. 2010). RNA isolated from these cells was examined by microarray analysis to identify mRNAs that were uniquely overexpressed in either the embryonic prostate stem cells or their niche. Two thousand four hundred eight transcripts were expressed with at least twofold higher levels in UGE than UGM, and 3,129 transcripts were expressed with at least twofold higher levels in UGM than UGE. From these differentially regulated genes, transcripts were selected that were expressed selectively in UGE cells that encode proteins that have cognate receptors or ligands expressed in UGM cells. In addition to pairs in which the ligand was expressed in one compartment and the receptor expressed in the other, indicating paracrine signaling, some ligand–receptor pairs were expressed in the same compartment, indicating autocrine signaling. Multiple ligand–receptor pairs were identified that may have roles in regulatory signaling between the stem cells and their niche (Blum et al. 2010). Prominent among the ligand–receptor pairs indicating paracrine or autocrine signaling were members of the BMP/TGF-ß/inhibin pathways. Other paracrine signaling pathways identified included the Wnt/ß-catenin, FGF, notch, hedgehog, PDGF, ephrin, and neurotrophic factor pathways. In a separate study examining molecules overexpressed in mouse adult prostate stem cells compared to differentiated cells, many of these same pathways were active in the adult stem cells, suggesting that these signaling molecules may also function in maintaining the adult niche (Blum et al. 2009). There is evidence that many of these pathways contribute to cell communication during prostate development.

### 6.5.1 TGF-ß

Members of the BMP/TGF-ß family are important regulators of stem cell quiescence and differentiation in embryonic, intestinal, mesenchymal, and skin stem cells (Watabe and Miyazono 2009). The TGF-ß signaling pathway also has an essential role in regulating prostate stem cell quiescence. The proximal region of the prostate responds differently to TGF-ß than the remainder of the gland (Tomlinson et al. 2004). High levels of TGF-ß signaling are found in the proximal region of normal adult prostate, and this signaling in combination with the high levels of Bcl-2 found in the proximal region presumably keeps the stem cells quiescent (Salm et al. 2005). This correlates with a signature of TGF-ß-induced genes that was observed in RNA expression analysis of adult prostatic stem cells (Blum et al. 2009). Androgen ablation increases TGF-ß signaling in the remainder of the prostate, where Bcl-2 levels are low, leading to apoptosis of cells in these regions. In contrast, androgen ablation decreases TGF-ß signaling in the proximal region, priming the stem cells to respond to growth factors when androgens are restored (Salm et al. 2005). Indeed, isolated prostatic stem cells produce abundant active

TGF-ß that inhibits their proliferation, differentiation into luminal cells, and ability to form ducts (Salm et al. 2012).

In addition to its importance in maintaining quiescence in the epithelium, TGF-ß signaling is also critical to the ability of the stroma to guide differentiation of stem cells. When UGM that was unresponsive to TGF-ß because of a fibroblast-specific knockout of TGF-ß receptor type II was co-inoculated with wild-type bladder epithelial cells under the renal capsule, the UGM was unable to transdifferentiate the bladder cells into prostate cells (Placencio et al. 2008; Li et al. 2009). Decreased TGF-ß signaling in the UGM affects stem cell function through alterations in Wnt/ß-catenin signals to the stem cells (Placencio et al. 2008; Li et al. 2009).

### *6.5.2 BMP*

Other members of the BMP/TGF-ß family that may alter stem cell function are BMP4 and BMP7. Although TGF-ß and the BMPs are structurally related, they interact with different receptors and activate a different set of signaling molecules. Both BMP4 and BMP7 are expressed in the prostatic mesenchyme during prostate development (Lamm et al. 2001; Grishina et al. 2005). Addition of BMP4 or BMP7 to urogenital sinuses cultured in vitro inhibited epithelial cell proliferation and decreased ductal branching (Lamm et al. 2001; Grishina et al. 2005). In contrast, adult BMP4 haploinsufficient mice and BMP7 knockout mice had increased ductal tips (Lamm et al. 2001; Grishina et al. 2005).

### *6.5.3 Wnt/β-Catenin*

The Wnt/ß-catenin pathway may also play a role in modulating prostate stem cell functions. The Wnt pathway regulates other stem cells, including keratinocyte, embryonal, colon, intestinal, and follicular stem cells (Reya and Clevers 2005; Wray and Hartmann 2012) and is critical in maintaining quiescence in hematopoietic stem cells (Fleming et al. 2008). The Wnt/ß-catenin pathway may be utilized by adult prostate stem cells, as a fraction of the Sca-1-positive cells (the population that contains stem cells) expressed a marker of activation of this pathway, axin2 (Ontiveros et al. 2008). The relevance of the Wnt/ß-catenin pathway to prostate stem cells is further indicated by the finding that knockout of secreted frizzled-related protein 1 (SFRP1), a homolog of the frizzled receptor for Wnt proteins that is expressed in prostatic stroma, inhibited epithelial proliferation in developing prostates, reduced branching, and increased expression of secretory proteins. Overexpression of SFRP1 had the opposite results, with increased epithelial proliferation and decreased expression of secretory proteins (Joesting et al. 2005, 2008). These results suggest that the stromal-derived SFRP1 regulates prostatic stem cell proliferation and differentiation.

### 6.5.4 FGF

The FGF family of growth factors has been implicated in the regulation of self-renewal in stem cells from a variety of organs (Coutu and Galipeau 2011; Eiselleova et al. 2009) and is essential for the maintenance of the embryonic stem cell niche (Bendall et al. 2007). The FGF ligands and receptors were also identified as potential niche signaling molecules in RNA array analysis of the stem cell niche (Blum et al. 2010). FGF10 is expressed in the prostatic mesenchyme and is required for the development of the prostate (Donjacour et al. 2003; Thomson and Cunha 1999). In FGF10 knockout mice, only rudimentary prostatic buds were occasionally observed in the urogenital sinus (Donjacour et al. 2003). Administration of FGF10 and testosterone to the rudimentary buds in culture partially restored prostate duct formation (Donjacour et al. 2003). Conditional knockout of FGF receptor-2, a receptor for FGF10, in prostatic epithelium ablated the development of the ventral and anterior lobes of the prostate and resulted in reduced growth of the dorsolateral lobes (Lin et al. 2007). Similarly, conditional knockout of FRS2α, an intracellular signaling molecule in the FGF pathway, in prostate epithelial cells resulted in reduced growth of the gland, although all lobes were present (Zhang et al. 2008). The discrepancy in results from knockout of the growth factor, its receptor, and their signaling intermediate may be due to the promiscuity of these molecules, in which the ligands can interact with multiple receptors and the receptors can signal through several intermediates. Nevertheless, the importance of FGF signaling in communication between prostate stem cells and their niche was emphasized by the observation that when UGM cells overexpressing FGF10 were co-inoculated under the renal capsule with prostatic stem cells, there was hyperproliferation in the epithelial compartment of the prostate-like organs formed (Memarzadeh et al. 2007).

### 6.5.5 Notch

The notch signaling pathway is also critical for a variety of tissue-specific stem cells (Bolos et al. 2007; Lai 2004). In prostates, this pathway is involved in both epithelial and stromal development. Administration of a γ-secretase inhibitor, which inhibits notch signaling, to explants of developing prostate increased epithelial cell proliferation and led to an accumulation of undifferentiated cells that expressed both luminal and basal cell markers (Wang et al. 2006b). Induced knockout of notch-1 in prostate epithelial cells had similar effects on prostates in situ (Wang et al. 2004, 2006b). In agreement with these results, expression of a constitutively active notch-1 receptor increased cell proliferation in the basal cell compartment of the epithelium and in smooth muscle (Wu et al. 2011). Knockout of a notch signaling molecule, Cbf1/Rbp-J, decreased cell proliferation in both compartments (Wu et al. 2011). The effects of notch on smooth muscle may be an example of cross signaling between cells in the stromal compartment. Notch-2 and a potential notch ligand, delta-like-1, are expressed in nonoverlapping regions of developing prostatic smooth muscle (Orr et al. 2009). Inhibition of notch signaling with γ-secretase inhibitor inhibits smooth muscle differentiation (Orr et al. 2009).

### 6.5.6 Hedgehog

Hedgehog signaling is important in regulating neural and hematopoietic stem cell proliferation, survival, and differentiation (Komada 2012; Vue et al. 2009; Bhardwaj et al. 2001). Hedgehog signaling may have a similar role in modulating prostate stem cells. Sonic hedgehog is expressed in the urogenital sinus epithelium during prostate development but becomes restricted to the proximal (stem cell containing) region of the developing ducts as prostate differentiation progresses (Podlasek et al. 1999). Patched, a hedgehog receptor that is induced in response to hedgehog signaling, is expressed in the UGM, suggesting epithelial to mesenchyme signaling (Podlasek et al. 1999; Lamm et al. 2002). Antibodies to sonic hedgehog blocked prostate development in embryonic prostates transplanted under the renal capsule of recipient mice and prostate regeneration in castrated mice administered testosterone (Podlasek et al. 1999; Karhadkar et al. 2004). However, others have found that addition of sonic hedgehog or activation of hedgehog signaling in prostate explants inhibits prostate ductal development (Pu et al. 2004; Freestone et al. 2003; Wang et al. 2003). These discrepancies may be due to the complex interaction of sonic hedgehog, its homolog Indian hedgehog, and the multiple signaling pathways they control (Doles et al. 2006). Nevertheless, it is clear that hedgehog signaling between the stroma and epithelium plays a critical role in prostate development.

### 6.5.7 Ephrins

There is evidence that ephrins and their receptors, the Eph proteins, play a role in the regulation of stem cells in a variety of organs. Ephrins and Eph receptors support early thymocyte survival and maturation (Alfaro et al. 2007; Stimamiglio et al. 2010; Wu and Luo 2005). Bidirectional signaling between ephrin B2 on osteoclasts and Eph B4 on osteoblasts regulates osteoclast maturation (Zhao et al. 2006). Knockdown of Eph A4 in neural stem cells caused premature differentiation, suggesting that this receptor maintains the stem cells in an undifferentiated state (Khodosevich et al. 2011). EphB2 expression identifies colonic stem cells (Jung et al. 2012), and administration of soluble forms of ephrin B2 or Eph B2 extracellular domain decreased cell proliferation in the intestinal crypts, suggesting that ephrin signaling regulates intestinal progenitor cell turnover (Holmberg et al. 2006). Disruption of either ephrin A or B interactions with Ephs in mouse skin doubled the rate of proliferation in hair follicles and epidermis, suggesting that these molecules are negative regulators of proliferation (Genander et al. 2010). Thus, ephrin–Eph interactions can modify the proliferation or differentiation of progenitor cells from a variety of organs, but their role in regulating prostate stem cells has not been addressed. The fact that four ephrins and eight Eph receptor homologs were identified as differentially expressed between embryonic prostate stem cells and the UGM may indicate that these molecules are also important in signaling in the prostate niche (Blum et al. 2010).

## 6.6 Aberrant Signaling from the Stroma Can Lead to Tumorigenesis

Disruption of the normal communication between stem cells and their niche can alter prostate homeostasis. It is obvious that decreased stimulatory signals or increased inhibitory signals may lead to atrophy of the gland as occurs after castration. However, there is evidence that increased stimulatory signals or decreased inhibitory signals can lead to tumor formation. Conditional knockout of the TGF-ß type II receptor in fibroblasts removed their growth inhibition by TGF-ß, caused an expansion of the stromal compartment, and induced PIN-type lesions in mouse prostate epithelium (Bhowmick et al. 2004). Presumably, the decreased TGF-ß regulation resulted in increased stromal growth stimulatory signaling to the stem cell compartment. Similarly, directly increasing growth stimulatory signals from the stroma by overexpression of FGF10, a stroma-derived epithelial growth factor, in UGM cells caused formation of well-differentiated prostate adenocarcinomas when these cells were co-transplanted with normal prostatic stem cells under the renal capsule of recipient mice (Memarzadeh et al. 2007). Tumor formation was abrogated if the stem cells were transfected with a dominant-negative FGF receptor-1 (Memarzadeh et al. 2007), showing that the stem cells were responding to the increased growth factor (Memarzadeh et al. 2007). Conversely, increased growth stimulatory signals from the stem cells to the stroma may also result in tumor formation. FGF8 is an epithelial-derived growth factor that normally signals to the stroma. Overexpression of FGF8 in prostate epithelium resulted in an atypical hypercellular stroma with an increased proportion of fibroblastic cells, rich vasculature, and inflammation (Song et al. 2002; Elo et al. 2012). Hyperplasia of the epithelium was also observed, leading to PIN lesions and sometimes progressing to adenocarcinoma in older animals (Song et al. 2002; Elo et al. 2012).

The formation of tumors as a result of imbalances in signaling between the niche and stem cells resembles the interaction of prostate tumor cells with their reactive stroma. It has been recognized that tumor stroma is fundamentally different from normal stroma and plays an important role in the aggressiveness of the tumor (Schor et al. 1987). In cell transplant experiments, initiated prostate epithelial cells from benign prostatic hyperplasia form tumors when co-inoculated with prostate tumor stromal cells but not when co-inoculated with stromal cells from benign prostatic hypertrophy or normal stromal cells, suggesting that tumor reactive stroma sends abnormal growth stimulatory signals to initiated epithelial stem cells (Olumi et al. 1999; Barclay et al. 2005) However, in contrast to the aberrant signaling between normal stem cells and stroma described above, combinations of tumor reactive stroma with normal epithelial cells did not lead to tumor formation (Olumi et al. 1999). This may be due to lower levels of these signaling molecules in the tumor reactive stroma. Thus, as tumors are thought to arise from transformation of normal organ stem cells, the reactive stroma may represent a niche for tumor stem cells or tumor-initiating cells. Some of the pathways used in tumor reactive stroma signaling to the tumor epithelium are the same as those identified for interactions of the embryonic niche with normal stem cells, including TGF-ß, Wnt/ß-catenin, and hedgehog pathways (Basanta et al. 2009; Franco et al. 2011; Ao et al. 2007; Li et al. 2008; Shaw et al. 2009).

## 6.7 Stem Cell Niche and Tumorigenesis

Normal stem cells have been proposed to be the source of tumor-initiating cells (Visvader and Lindeman 2012; Tang 2012). In this view, oncogenic transformation of a normal stem cell leads to the generation of tumor-initiating cells that undergo uncontrolled proliferation. Some tumor-initiating cells differentiate, forming the mixture of cells normally found in a tumor, while others maintain stemlike properties, including a high proliferative potential. Prostate tumors seem to follow this model. Prostate tumors in mice were shown to arise from the stem cell-containing proximal region of the ducts (Zhou et al. 2007). Transformation of both mouse and human prostate stem cells led to formation of adenocarcinomas, whereas transformation of non-stem cell populations did not result in tumors (Lawson et al. 2010; Goldstein et al. 2010, 2011). As noted before, stem cells have been identified in both the basal and luminal epithelial compartments (Goldstein et al. 2010; Lawson et al. 2010; Wang et al. 2009). Transformation of either basal or luminal stem cells has been shown to form tumors (Wang et al. 2009; Goldstein et al. 2010; Lawson et al. 2010; Choi et al. 2012; Moscatelli and Wilson 2010). Transformation leads to an expansion of cells with stem cell markers (Wang et al. 2006a). Once tumors arise, the stroma appears to be altered along with the tumor, suggesting that the niche is expanded to support the growth of the tumor-initiating cells (Sung and Chung 2002; Josson et al. 2010). Even prostate tumor metastases are dependent on the support of a niche. Metastases to bone co-opt the hematopoietic stem cell niche in the bone marrow, where they take on some of the properties of hematopoietic stem cells, including the ability to be mobilized into the circulation by administration of GM-CSF (Shiozawa et al. 2011). Thus, formation of tumors, their expansion, and their metastatic spread are modulated by interactions with the niche.

## 6.8 Summary

The prostate stem cell niche consists of a localized area of the proximal region of the ducts that contains a high concentration of epithelial stem cells. The stem cells interact with other epithelial cells and underlying stromal cells to modulate their quiescence, proliferation, and differentiation (Fig. 6.1). Several signaling molecules involved in these interactions have been identified, and others have been proposed. Some of these signals are paracrine, with molecules produced by the epithelium acting on the stroma and molecules produced by the stroma acting on the epithelium (Fig. 6.1). Other signals are autocrine, with the signaling molecule produced by and affecting the same cell type, or paracrine within a compartment, for example, one stromal cell type signaling to another. However, the specific signals and their cellular sources and targets have only begun to be elucidated. Additional investigations with cell-specific knockout or overexpression of the signaling molecules and their receptors are necessary to map out the complex interactions. Knowledge of these interactions will provide insights into the etiology of proliferative diseases of the prostate, both benign and malignant.

**Acknowledgements** This work was supported by the National Institutes of Health (CA132641 (ELW), AG34305 (DM)), the Helen L. and Martin S. Kimmel Center for Stem Cell Biology, and the New York University School of Medicine NIH/NCI Cancer Center Support Grant (5 P30 CA016087032).

## References

Alfaro D, Garcia-Ceca JJ, Cejalvo T, Jimenez E, Jenkinson EJ, Anderson G, Munoz JJ, Zapata A (2007) EphrinB1-EphB signaling regulates thymocyte-epithelium interactions involved in functional T cell development. Eur J Immunol 37:2596–2605

Ao M, Franco OE, Park D, Raman D, Williams K, Hayward SW (2007) Cross-talk between paracrine-acting cytokine and chemokine pathways promotes malignancy in benign human prostatic epithelium. Cancer Res 67:4244–4253

Barclay WW, Woodruff RD, Hall MC, Cramer SD (2005) A system for studying epithelial-stromal interactions reveals distinct inductive abilities of stromal cells from benign prostatic hyperplasia and prostate cancer. Endocrinology 146:13–18

Basanta D, Strand DW, Lukner RB, Franco OE, Cliffel DE, Ayala GE, Hayward SW, Anderson AR (2009) The role of transforming growth factor-beta-mediated tumor-stroma interactions in prostate cancer progression: an integrative approach. Cancer Res 69:7111–7120

Bates M, Kovalenko B, Wilson EL, Moscatelli D (2008) Endothelial cells support the growth of prostate tissue in vivo. Prostate 68:893–901

Bendall SC, Stewart MH, Menendez P, George D, Vijayaragavan K, Werbowetski-Ogilvie T, Ramos-Mejia V, Rouleau A, Yang J, Bosse M, Lajoie G, Bhatia M (2007) IGF and FGF cooperatively establish the regulatory stem cell niche of pluripotent human cells in vitro. Nature 448:1015–1021

Bhardwaj G, Murdoch B, Wu D, Baker DP, Williams KP, Chadwick K, Ling LE, Karanu FN, Bhatia M (2001) Sonic hedgehog induces the proliferation of primitive human hematopoietic cells via BMP regulation. Nat Immunol 2:172–180

Bhowmick NA, Chytil A, Plieth D, Gorska AE, Dumont N, Shappell S, Washington MK, Neilson EG, Moses HL (2004) TGF-beta signaling in fibroblasts modulates the oncogenic potential of adjacent epithelia. Science 303:848–851

Blum R, Gupta R, Burger PE, Ontiveros CS, Salm SN, Xiong X, Kamb A, Wesche H, Marshall L, Cutler G, Wang X, Zavadil J, Moscatelli D, Wilson EL (2009) Molecular signatures of prostate stem cells reveal novel signaling pathways and provide insights into prostate cancer. PLoS One 4:e5722

Blum R, Gupta R, Burger PE, Ontiveros CS, Salm SN, Xiong X, Kamb A, Wesche H, Marshall L, Cutler G, Wang X, Zavadil J, Moscatelli D, Wilson EL (2010) Molecular signatures of the primitive prostate stem cell niche reveal novel mesenchymal-epithelial signaling pathways. PLoS One 5:e13024. doi:10.1371/journal.pone.0013024

Bolos V, Grego-Bessa J, de la Pompa JL (2007) Notch signaling in development and cancer. Endocr Rev 28:339–363

Boutin EL, Battle E, Cunha GR (1991) The response of female urogenital tract epithelia to mesenchymal inductors is restricted by the germ layer origin of the epithelium: prostatic inductions. Differentiation 48:99–105

Burger PE, Xiong X, Coetzee S, Salm SN, Moscatelli D, Goto K, Wilson EL (2005) Sca-1 expression identifies stem cells in the proximal region of prostatic ducts with high capacity to reconstitute prostatic tissue. Proc Natl Acad Sci USA 102:7180–7185

Chan RW, Gargett CE (2006) Identification of label-retaining cells in mouse endometrium. Stem Cells 24:1529–1538

Choi N, Zhang B, Zhang L, Ittmann M, Xin L (2012) Adult murine prostate basal and luminal cells are self-sustained lineages that can both serve as targets for prostate cancer initiation. Cancer Cell 21:253–265

Chung LW, Cunha GR (1983) Stromal-epithelial interactions: II. Regulation of prostatic growth by embryonic urogenital sinus mesenchyme. Prostate 4:503–511

Clevers H (2009) Searching for adult stem cells in the intestine. EMBO Mol Med 1:255–259

Coutu DL, Galipeau J (2011) Roles of FGF signaling in stem cell self-renewal, senescence and aging. Aging (Albany NY) 3:920–933

Cunha GR, Lung B, Reese B (1980) Glandular epithelial induction by embryonic mesenchyme in adult bladder epithelium of BALB/c mice. Invest Urol 17:302–304

Cunha GR, Fujii H, Neubauer BL, Shannon JM, Sawyer L, Reese BA (1983) Epithelial-mesenchymal interactions in prostatic development. I. Morphological observations of prostatic induction by urogenital sinus mesenchyme in epithelium of the adult rodent urinary bladder. J Cell Biol 96:1662–1670

Cunha GR, Hayward SW, Dahiya R, Foster BA (1996) Smooth muscle-epithelial interactions in normal and neoplastic prostatic development. Acta Anat (Basel) 155:63–72

Ding L, Saunders TL, Enikolopov G, Morrison SJ (2012) Endothelial and perivascular cells maintain haematopoietic stem cells. Nature 481:457–462

Doles J, Cook C, Shi X, Valosky J, Lipinski R, Bushman W (2006) Functional compensation in Hedgehog signaling during mouse prostate development. Dev Biol 295:13–25

Donjacour AA, Thomson AA, Cunha GR (2003) FGF-10 plays an essential role in the growth of the fetal prostate. Dev Biol 261:39–54

Eiselleova L, Matulka K, Kriz V, Kunova M, Schmidtova Z, Neradil J, Tichy B, Dvorakova D, Pospisilova S, Hampl A, Dvorak P (2009) A complex role for FGF-2 in self-renewal, survival, and adhesion of human embryonic stem cells. Stem Cells 27:1847–1857

Elo T, Sipila P, Valve E, Kujala P, Toppari J, Poutanen M, Harkonen P (2012) Fibroblast growth factor 8b causes progressive stromal and epithelial changes in the epididymis and degeneration of the seminiferous epithelium in the testis of transgenic mice. Biol Reprod 86(157):1–12

English HF, Drago JR, Santen RJ (1985) Cellular response to androgen depletion and repletion in the rat ventral prostate: autoradiography and morphometric analysis. Prostate 7:41–51

Fleming HE, Janzen V, Lo Celso C, Guo J, Leahy KM, Kronenberg HM, Scadden DT (2008) Wnt signaling in the niche enforces hematopoietic stem cell quiescence and is necessary to preserve self-renewal in vivo. Cell Stem Cell 2:274–283

Folkman J (1998) Is tissue mass regulated by vascular endothelial cells? Prostate as the first evidence. Endocrinology 139:441–442

Franck-Lissbrant I, Haggstrom S, Damber JE, Bergh A (1998) Testosterone stimulates angiogenesis and vascular regrowth in the ventral prostate in castrated adult rats. Endocrinology 139:451–456

Franco OE, Jiang M, Strand DW, Peacock J, Fernandez S, Jackson RS 2nd, Revelo MP, Bhowmick NA, Hayward SW (2011) Altered TGF-beta signaling in a subpopulation of human stromal cells promotes prostatic carcinogenesis. Cancer Res 71:1272–1281

Freestone SH, Marker P, Grace OC, Tomlinson DC, Cunha GR, Harnden P, Thomson AA (2003) Sonic hedgehog regulates prostatic growth and epithelial differentiation. Dev Biol 264:352–362

Fuchs E, Horsley V (2008) More than one way to skin. Genes Dev 22:976–985

Genander M, Holmberg J, Frisen J (2010) Ephrins negatively regulate cell proliferation in the epidermis and hair follicle. Stem Cells 28:1196–1205

Goldstein AS, Lawson DA, Cheng D, Sun W, Garraway IP, Witte ON (2008) Trop2 identifies a subpopulation of murine and human prostate basal cells with stem cell characteristics. Proc Natl Acad Sci USA 105:20882–20887

Goldstein AS, Huang J, Guo C, Garraway IP, Witte ON (2010) Identification of a cell of origin for human prostate cancer. Science 329:568–571

Goldstein AS, Drake JM, Burnes DL, Finley DS, Zhang H, Reiter RE, Huang J, Witte ON (2011) Purification and direct transformation of epithelial progenitor cells from primary human prostate. Nat Protoc 6:656–667

Goto K, Salm SN, Coetzee S, Xiong X, Burger PE, Shapiro E, Lepor H, Moscatelli D, Wilson EL (2006) Proximal prostatic stem cells are programmed to regenerate a proximal-distal ductal axis. Stem Cells 24:1859–1868

Grishina IB, Kim SY, Ferrara C, Makarenkova HP, Walden PD (2005) BMP7 inhibits branching morphogenesis in the prostate gland and interferes with Notch signaling. Dev Biol 288:334–347

Hayward SW, Haughney PC, Rosen MA, Greulich KM, Weier HU, Dahiya R, Cunha GR (1998) Interactions between adult human prostatic epithelium and rat urogenital sinus mesenchyme in a tissue recombination model. Differentiation 63:131–140

Hayward SW, Haughney PC, Lopes ES, Danielpour D, Cunha GR (1999) The rat prostatic epithelial cell line NRP-152 can differentiate in vivo in response to its stromal environment. Prostate 39:205–212

Holmberg J, Genander M, Halford MM, Anneren C, Sondell M, Chumley MJ, Silvany RE, Henkemeyer M, Frisen J (2006) EphB receptors coordinate migration and proliferation in the intestinal stem cell niche. Cell 125:1151–1163

Joesting MS, Perrin S, Elenbaas B, Fawell SE, Rubin JS, Franco OE, Hayward SW, Cunha GR, Marker PC (2005) Identification of SFRP1 as a candidate mediator of stromal-to-epithelial signaling in prostate cancer. Cancer Res 65:10423–10430

Joesting MS, Cheever TR, Volzing KG, Yamaguchi TP, Wolf V, Naf D, Rubin JS, Marker PC (2008) Secreted frizzled related protein 1 is a paracrine modulator of epithelial branching morphogenesis, proliferation, and secretory gene expression in the prostate. Dev Biol 317:161–173

Josson S, Matsuoka Y, Chung LW, Zhau HE, Wang R (2010) Tumor-stroma co-evolution in prostate cancer progression and metastasis. Semin Cell Dev Biol 21:26–32

Jung P, Sato T, Merlos-Suarez A, Barriga FM, Iglesias M, Rossell D, Auer H, Gallardo M, Blasco MA, Sancho E, Clevers H, Batlle E (2012) Isolation and in vitro expansion of human colonic stem cells. Nat Med 17:1225–1227

Karhadkar SS, Bova GS, Abdallah N, Dhara S, Gardner D, Maitra A, Isaacs JT, Berman DM, Beachy PA (2004) Hedgehog signalling in prostate regeneration, neoplasia and metastasis. Nature 431:707–712

Khodosevich K, Watanabe Y, Monyer H (2011) EphA4 preserves postnatal and adult neural stem cells in an undifferentiated state in vivo. J Cell Sci 124:1268–1279

Kiel MJ, Yilmaz OH, Iwashita T, Terhorst C, Morrison SJ (2005) SLAM family receptors distinguish hematopoietic stem and progenitor cells and reveal endothelial niches for stem cells. Cell 121:1109–1121

Komada M (2012) Sonic hedgehog signaling coordinates the proliferation and differentiation of neural stem/progenitor cells by regulating cell cycle kinetics during development of the neocortex. Congenit Anom (Kyoto) 52:72–77

Kunisaki Y, Frenette PS (2012) The secrets of the bone marrow niche: enigmatic niche brings challenge for HSC expansion. Nat Med 18:864–865

Lai EC (2004) Notch signaling: control of cell communication and cell fate. Development 131:965–973

Lamm ML, Podlasek CA, Barnett DH, Lee J, Clemens JQ, Hebner CM, Bushman W (2001) Mesenchymal factor bone morphogenetic protein 4 restricts ductal budding and branching morphogenesis in the developing prostate. Dev Biol 232:301–314

Lamm ML, Catbagan WS, Laciak RJ, Barnett DH, Hebner CM, Gaffield W, Walterhouse D, Iannaccone P, Bushman W (2002) Sonic hedgehog activates mesenchymal Gli1 expression during prostate ductal bud formation. Dev Biol 249:349–366

Lavker RM, Tseng SC, Sun TT (2004) Corneal epithelial stem cells at the limbus: looking at some old problems from a new angle. Exp Eye Res 78:433–446

Lawson DA, Zong Y, Memarzadeh S, Xin L, Huang J, Witte ON (2010) Basal epithelial stem cells are efficient targets for prostate cancer initiation. Proc Natl Acad Sci USA 107:2610–2615

Li X, Placencio V, Iturregui JM, Uwamariya C, Sharif-Afshar AR, Koyama T, Hayward SW, Bhowmick NA (2008) Prostate tumor progression is mediated by a paracrine TGF-beta/Wnt3a signaling axis. Oncogene 27:7118–7130

Li X, Wang Y, Sharif-Afshar AR, Uwamariya C, Yi A, Ishii K, Hayward SW, Matusik RJ, Bhowmick NA (2009) Urothelial transdifferentiation to prostate epithelia is mediated by paracrine TGF-beta signaling. Differentiation 77:95–102

Lin Y, Liu G, Zhang Y, Hu YP, Yu K, Lin C, McKeehan K, Xuan JW, Ornitz DM, Shen MM, Greenberg N, McKeehan WL, Wang F (2007) Fibroblast growth factor receptor 2 tyrosine kinase is required for prostatic morphogenesis and the acquisition of strict androgen dependency for adult tissue homeostasis. Development 134:723–734

Lissbrant IF, Hammarsten P, Lissbrant E, Ferrara N, Rudolfsson SH, Bergh A (2004) Neutralizing VEGF bioactivity with a soluble chimeric VEGF-receptor protein flt(1–3)IgG inhibits testosterone-stimulated prostate growth in castrated mice. Prostate 58:57–65

Memarzadeh S, Xin L, Mulholland DJ, Mansukhani A, Wu H, Teitell MA, Witte ON (2007) Enhanced paracrine FGF10 expression promotes formation of multifocal prostate adenocarcinoma and an increase in epithelial androgen receptor. Cancer Cell 12:572–585

Miller FD, Gauthier-Fisher A (2009) Home at last: neural stem cell niches defined. Cell Stem Cell 4:507–510

Moscatelli D, Wilson EL (2010) PINing down the origin of prostate cancer. Sci Transl Med 2:43ps38

Nemeth JA, Lee C (1996) Prostatic ductal system in rats: regional variation in stromal organization. Prostate 28:124–128

Neubauer BL, Chung LW, McCormick KA, Taguchi O, Thompson TC, Cunha GR (1983) Epithelial-mesenchymal interactions in prostatic development. II. Biochemical observations of prostatic induction by urogenital sinus mesenchyme in epithelium of the adult rodent urinary bladder. J Cell Biol 96:1671–1676

Norman JT, Cunha GR, Sugimura Y (1986) The induction of new ductal growth in adult prostatic epithelium in response to an embryonic prostatic inductor. Prostate 8:209–220

Ohab JJ, Fleming S, Blesch A, Carmichael ST (2006) A neurovascular niche for neurogenesis after stroke. J Neurosci 26:13007–13016

Olumi AF, Grossfeld GD, Hayward SW, Carroll PR, Tlsty TD, Cunha GR (1999) Carcinoma-associated fibroblasts direct tumor progression of initiated human prostatic epithelium. Cancer Res 59:5002–5011

Ontiveros CS, Salm SN, Wilson EL (2008) Axin2 expression identifies progenitor cells in the murine prostate. Prostate 68:1263–1272

Orford KW, Scadden DT (2008) Deconstructing stem cell self-renewal: genetic insights into cell-cycle regulation. Nat Rev Genet 9:115–128

Orr B, Grace OC, Vanpoucke G, Ashley GR, Thomson AA (2009) A role for notch signaling in stromal survival and differentiation during prostate development. Endocrinology 150:463–472

Palmer TD, Willhoite AR, Gage FH (2000) Vascular niche for adult hippocampal neurogenesis. J Comp Neurol 425:479–494

Placencio VR, Sharif-Afshar AR, Li X, Huang H, Uwamariya C, Neilson EG, Shen MM, Matusik RJ, Hayward SW, Bhowmick NA (2008) Stromal transforming growth factor-beta signaling mediates prostatic response to androgen ablation by paracrine Wnt activity. Cancer Res 68:4709–4718

Podlasek CA, Barnett DH, Clemens JQ, Bak PM, Bushman W (1999) Prostate development requires Sonic hedgehog expressed by the urogenital sinus epithelium. Dev Biol 209:28–39

Pu Y, Huang L, Prins GS (2004) Sonic hedgehog-patched Gli signaling in the developing rat prostate gland: lobe-specific suppression by neonatal estrogens reduces ductal growth and branching. Dev Biol 273:257–275

Reya T, Clevers H (2005) Wnt signalling in stem cells and cancer. Nature 434:843–850

Richardson GD, Robson CN, Lang SH, Neal DE, Maitland NJ, Collins AT (2004) CD133, a novel marker for human prostatic epithelial stem cells. J Cell Sci 117:3539–3545

Salm SN, Burger PE, Coetzee S, Goto K, Moscatelli D, Wilson EL (2005) TGF-{beta} maintains dormancy of prostatic stem cells in the proximal region of ducts. J Cell Biol 170:81–90

Salm S, Burger PE, Wilson EL (2012) TGF-beta and stem cell factor regulate cell proliferation in the proximal stem cell niche. Prostate 72:998–1005

Schor SL, Schor AM, Howell A, Crowther D (1987) Hypothesis: persistent expression of fetal phenotypic characteristics by fibroblasts is associated with an increased susceptibility to neoplastic disease. Exp Cell Biol 55:11–17

Shabisgh A, Tanji N, D'Agati V, Burchardt M, Rubin M, Goluboff ET, Heitjan D, Kiss A, Buttyan R (1999) Early effects of castration on the vascular system of the rat ventral prostate gland. Endocrinology 140:1920–1926

Shaw A, Gipp J, Bushman W (2009) The Sonic hedgehog pathway stimulates prostate tumor growth by paracrine signaling and recapitulates embryonic gene expression in tumor myofibroblasts. Oncogene 28:4480–4490

Shi S, Gronthos S (2003) Perivascular niche of postnatal mesenchymal stem cells in human bone marrow and dental pulp. J Bone Miner Res 18:696–704

Shiozawa Y, Pedersen EA, Havens AM, Jung Y, Mishra A, Joseph J, Kim JK, Patel LR, Ying C, Ziegler AM, Pienta MJ, Song J, Wang J, Loberg RD, Krebsbach PH, Pienta KJ, Taichman RS (2011) Human prostate cancer metastases target the hematopoietic stem cell niche to establish footholds in mouse bone marrow. J Clin Invest 121:1298–1312

Song Z, Wu X, Powell WC, Cardiff RD, Cohen MB, Tin RT, Matusik RJ, Miller GJ, Roy-Burman P (2002) Fibroblast growth factor 8 isoform B overexpression in prostate epithelium: a new mouse model for prostatic intraepithelial neoplasia. Cancer Res 62:5096–5105

Stimamiglio MA, Jimenez E, Silva-Barbosa SD, Alfaro D, Garcia-Ceca JJ, Munoz JJ, Cejalvo T, Savino W, Zapata A (2010) EphB2-mediated interactions are essential for proper migration of T cell progenitors during fetal thymus colonization. J Leukoc Biol 88:483–494

Sung SY, Chung LW (2002) Prostate tumor-stroma interaction: molecular mechanisms and opportunities for therapeutic targeting. Differentiation 70:506–521

Takao T, Tsujimura A, Coetzee S, Salm SN, Lepor H, Shapiro E, Moscatelli D, Wilson EL (2003) Stromal/epithelial interactions of murine prostatic cell lines in vivo: a model for benign prostatic hyperplasia and the effect of doxazosin on tissue size. Prostate 54:17–24
Tang DG (2012) Understanding cancer stem cell heterogeneity and plasticity. Cell Res 22:457–472
Thomson AA, Cunha GR (1999) Prostatic growth and development are regulated by FGF10. Development 126:3693–3701
Tokuda Y, Toda S, Masaki Z, Sugihara H (1999) Proliferation and differentiation of rat dorsal prostatic epithelial cells in collagen gel matrix culture, focusing upon effects of adipocytes. Int J Urol 6:509–519
Tomlinson DC, Freestone SH, Grace OC, Thomson AA (2004) Differential effects of transforming growth factor-beta1 on cellular proliferation in the developing prostate. Endocrinology 145:4292–4300
Tsujimura A, Koikawa Y, Salm S, Takao T, Coetzee S, Moscatelli D, Shapiro E, Lepor H, Sun TT, Wilson EL (2002) Proximal location of mouse prostate epithelial stem cells: a model of prostatic homeostasis. J Cell Biol 157:1257–1265
Visvader JE, Lindeman GJ (2012) Cancer stem cells: current status and evolving complexities. Cell Stem Cell 10:717–728
Vue TY, Bluske K, Alishahi A, Yang LL, Koyano-Nakagawa N, Novitch B, Nakagawa Y (2009) Sonic hedgehog signaling controls thalamic progenitor identity and nuclei specification in mice. J Neurosci 29:4484–4497
Wang BE, Shou J, Ross S, Koeppen H, De Sauvage FJ, Gao WQ (2003) Inhibition of epithelial ductal branching in the prostate by sonic hedgehog is indirectly mediated by stromal cells. J Biol Chem 278:18506–18513
Wang XD, Shou J, Wong P, French DM, Gao WQ (2004) Notch1-expressing cells are indispensable for prostatic branching morphogenesis during development and re-growth following castration and androgen replacement. J Biol Chem 279:24733–24744
Wang S, Garcia AJ, Wu M, Lawson DA, Witte ON, Wu H (2006a) Pten deletion leads to the expansion of a prostatic stem/progenitor cell subpopulation and tumor initiation. Proc Natl Acad Sci USA 103:1480–1485
Wang XD, Leow CC, Zha J, Tang Z, Modrusan Z, Radtke F, Aguet M, de Sauvage FJ, Gao WQ (2006b) Notch signaling is required for normal prostatic epithelial cell proliferation and differentiation. Dev Biol 290:66–80
Wang G, Kovalenko B, Huang Y, Moscatelli D (2007a) Vascular endothelial growth factor and angiopoietin are required for prostate regeneration. Prostate 67:485–499
Wang GM, Kovalenko B, Wilson EL, Moscatelli D (2007b) Vascular density is highest in the proximal region of the mouse prostate. Prostate 67:968–975
Wang X, Kruithof-de Julio M, Economides KD, Walker D, Yu H, Halili MV, Hu YP, Price SM, Abate-Shen C, Shen MM (2009) A luminal epithelial stem cell that is a cell of origin for prostate cancer. Nature 461:495–500
Watabe T, Miyazono K (2009) Roles of TGF-beta family signaling in stem cell renewal and differentiation. Cell Res 19:103–115
Wray J, Hartmann C (2012) WNTing embryonic stem cells. Trends Cell Biol 22:159–168
Wu J, Luo H (2005) Recent advances on T-cell regulation by receptor tyrosine kinases. Curr Opin Hematol 12:292–297
Wu X, Xu K, Zhang L, Deng Y, Lee P, Shapiro E, Monaco M, Makarenkova HP, Li J, Lepor H, Grishina I (2011) Differentiation of the ductal epithelium and smooth muscle in the prostate gland are regulated by the Notch/PTEN-dependent mechanism. Dev Biol 356:337–349
Xin L, Ide H, Kim Y, Dubey P, Witte ON (2003) In vivo regeneration of murine prostate from dissociated cell populations of postnatal epithelia and urogenital sinus mesenchyme. Proc Natl Acad Sci USA 100(Suppl 1):11896–11903
Xin L, Lawson DA, Witte ON (2005) The Sca-1 cell surface marker enriches for a prostate-regenerating cell subpopulation that can initiate prostate tumorigenesis. Proc Natl Acad Sci USA 102:6942–6947

Yamazaki S, Ema H, Karlsson G, Yamaguchi T, Miyoshi H, Shioda S, Taketo MM, Karlsson S, Iwama A, Nakauchi H (2011) Nonmyelinating Schwann cells maintain hematopoietic stem cell hibernation in the bone marrow niche. Cell 147:1146–1158

Yin T, Li L (2006) The stem cell niches in bone. J Clin Invest 116:1195–1201

Yoshida S, Sukeno M, Nabeshima Y (2007) A vasculature-associated niche for undifferentiated spermatogonia in the mouse testis. Science 317:1722–1726

Zhang Y, Zhang J, Lin Y, Lan Y, Lin C, Xuan JW, Shen MM, McKeehan WL, Greenberg NM, Wang F (2008) Role of epithelial cell fibroblast growth factor receptor substrate 2alpha in prostate development, regeneration and tumorigenesis. Development 135:775–784

Zhao C, Irie N, Takada Y, Shimoda K, Miyamoto T, Nishiwaki T, Suda T, Matsuo K (2006) Bidirectional ephrinB2-EphB4 signaling controls bone homeostasis. Cell Metab 4:111–121

Zhou Z, Flesken-Nikitin A, Nikitin AY (2007) Prostate cancer associated with p53 and Rb deficiency arises from the stem/progenitor cell-enriched proximal region of prostatic ducts. Cancer Res 67:5683–5690

# Chapter 7
# Tumour Stroma Control of Human Prostate Cancer Stem Cells

**Gail P. Risbridger and Renea A. Taylor**

**Abstract** Prostate cancer is an epithelial malignancy and stem cells are a major focus of current research efforts. This is a warranted approach, since the processes of self-renewal and differentiation underpin the fundamental biology of malignancy and cancer recurrence due to therapeutic resistance. In this chapter, we review the regulation of stem cells through reciprocal interaction with the surrounding microenvironment. In both normal tissues and in prostate cancer, stem cells are controlled by both intrinsic and extrinsic mechanisms, the latter involving stromal directed stem cell differentiation. Herein, we discuss the current experimental models to study stromal-stem cell interaction and present the current knowledge on how the two cellular compartments should be considered in unison to design more effective therapies for clinical management of prostate cancer.

## 7.1 Introduction

The focus of this book is on stem cells in prostate cancer. As in many systems, regulation of prostatic stem cells requires intrinsic and extrinsic mechanisms. This chapter will focus specifically on the extrinsic regulation, which has proven to be as important as intrinsic mechanisms of modulating stem cells. The multiple cell types in the prostate are located in epithelial or stromal compartments, each with their own role in maintaining structural architecture, secretory activity or differentiation. These cells work together in an orchestrated fashion to maintain glandular homeostasis. The stroma surrounding the epithelium regulates secretory function, as well

G.P. Risbridger, Ph.D. (✉) • R.A. Taylor, Ph.D.
Prostate Cancer Research Group, Department of Anatomy and Developmental Biology, School of Biomedical Sciences, Monash University, Level 3, Building 76, Wellington Road, Clayton, VIC 3168, Australia
e-mail: gail.risbridger@monash.edu

S.D. Cramer (ed.), *Stem Cells and Prostate Cancer*, DOI 10.1007/978-1-4614-6498-3_7, 

as differentiation of stem cells involved in initiation and progression of cells of origin of prostate cancer.

The identification of prostatic stem cells in normal and cancerous tissues is an obvious research priority, and significant advances have been made by several groups in the last 5 years. Yet the function and differentiation of stem cells by their stromal niche has also emerged as being important to understanding the extrinsic regulation of stem cells. In this chapter, we will review the literature on stromal requirement for prostate stem cell viability in normal tissues and in cancer, particularly focussing on in vivo models, and the use of stroma to direct differentiation and malignant transformation of stem cells in prostate cancer. Although the stroma is often peripheral to the discussion and perceptions about cancer stem cells or cells of origin of prostate cancer, it is a matter of fact that all of the assays used to detect stem cells in vivo required stoma. The necessity of stroma indicates it is essential and should be prominent in consideration of stem cell biology and regulatory mechanisms: the stem cells do not function in a vacuum and therefore cannot be considered in isolation. The goal of this chapter is to provide provocative opinion on the major role played by stroma in regulating prostate cancer stem cell function and stimulate this debate.

## 7.2 Defining Prostatic Stem Cells

There have been multiple seminal studies published in the field that identified and/or characterised prostatic stem cells in normal and tumour tissues. Most of these studies were in mice, but more recently included the use of human tissues. The discovery of cells in normal tissue, that have properties of long-term self-renewal and multi-lineage differentiation, underpins our fundamental understanding of the prostatic lineage and hierarchical arrangements, and more recently, their role in tumour initiation and progression (Taylor et al. 2012).

For clarity, in this chapter, we define the terms we use to describe different types of stem cells. Normal tissue stem cells are those that undergo self-renewal and repopulate the normal epithelial cells. Secondly, cells of origin are the cells in normal tissue (*can be stem cells or non-stem cells*) that have malignant potential and can give rise to tumours. Thirdly, cancer-repopulating (or stem) cells are the cells within a cancer that are hypothesised to self-renew and repopulate the bulk of a tumour. These definitions are stated in order to avoid the confusion of the term 'cancer- or tumour-initiating cells' which is not specific to either disease initiation or progression (Taylor et al. 2010).

### *7.2.1 Role of Stroma in Identifying Normal Prostatic Stem Cells*

In general, the approach to identifying prostate stem cells has been to use cell surface markers to isolate populations of cells that have stem cell properties, and in mouse, these have included $Lin^{-}CD49f^{hi}Sca\text{-}1^{hi}$ or $Lin^{-}Sca\text{-}1^{+}CD133^{+}CD44^{+}CD117^{+}$

and in human can include stem-enriched $\alpha2\beta1$integrin$^{hi}$/CD44$^{+}$/CD133$^{+}$ and basal Lin$^{-}$,CD49f$^{hi}$,Trop2$^{hi}$ cells (Goldstein et al. 2008, 2010; Lawson et al. 2007; Leong et al. 2008; Richardson et al. 2004; Xin et al. 2005; Burger et al. 2005), *discussed in more detail in other chapters of this book*. All of these cell fractions were defined by basal cell phenotypes, confirming the original prediction-based castration and testosterone replacement studies (English et al. 1987) that basal cells harbour prostatic stem cells. More recently, additional populations of stem cells were identified in the luminal compartment, initiated by the discovery of rare cells, visualised as castrate-resistant Nkx3.1-expressing cells (CARNs) that have multi-lineage and malignant potential (Wang et al. 2009). An alternate approach included in vivo tracing of lineage commitment and differentiation in normal prostate epithelia to determine the source of basal and luminal epithelia (Choi et al. 2012; Liu et al. 2011; Blackwood et al. 2011). In normal tissues, PSA-expressing luminal cells resist castration and regenerate luminal epithelium (Liu et al. 2011), and the same is true in cancer, although the PSA$^{lo}$-expressing cells harbour highly tumourigenic castration-resistant cancer cells (Qin et al. 2012).

This collection of high-profile publications significantly advanced the field, but compared to progress in other solid tissues such as breast and colon, our understanding in this area is comparably slow. For example, the study by Leong et al., showing CD117 was a unique prostatic stem cell surface marker, was the only study to use single cell transfer of fractionated murine epithelia (Leong et al. 2008); these cells have not yet been identified in human prostate. Secondly, the two cell fractions isolated by the Witte laboratory (LSC in mice and CD49$^{hi}$/Trop2$^{hi}$ in human) select for basal cells with enriched stem cell activity in vitro and in vivo. However, these are heterogeneous enriched populations of cells, and the full complement of prostatic epithelia cannot be derived from a single cell (Lukacs et al. 2010). The identification of luminal stem cells by the Shen Laboratory presents technical difficulties as CARNs can only be visualised in tissue following castration, using intracellular antibodies that limit their utility in sorting procedures (Wang et al. 2009). Nonetheless, enriched stem cell populations exist in the adult prostate, and they can have multiple phenotypes, being either basal or luminal cells, and their contribution to tissue repair and regeneration is vital.

Of all of these studies, the three main assays used to define stem cell potential have been (1) colony and sphere forming assays, (2) in vivo regeneration of prostatic epithelium and (3) castration and testosterone replacement. Whilst these have been adapted from other tissue types, their use in prostate biology is complex and warrants discussion.

The sphere forming assay involves aggregation of fractionated prostatic cells in a Matrigel mixture to which growth factors are added. This is an in vitro assay that has been widely used to retrospectively or prospectively identify stem cells based on the self-renewal and differentiation capabilities of single cells in normal and tumour tissues (Xin et al. 2007; Miki et al. 2007; Hu et al. 2011a, b; Goodyear et al. 2009). Whilst it provides a 3D model system that is supported by an enriched matrix of structural proteins, limitations include the ability of progenitor cells to generate spheres for short periods of time, lack of stromal cell interaction and failure to undergo full differentiation into a mature pseudostratified glandular epithelium.

Therefore, the proof of stem cell identity relies heavily on the use of in vivo transplantation studies.

In vivo confirmation of stem cell activity for multiple cell fractions, including all of those mentioned above, involves transplantation of single or enriched populations of cells, often co-grafted with stromal cells, into immune-deficient host mice to determine the potential to differentiate and generate the pseudostratified prostatic epithelium (Xin et al. 2003). In similar studies to identify mammary stem cells, the cleared mammary fat pad model was used where single cells were implanted and full outgrowths generated (Shackleton et al. 2006), but an equivalent bioassay in prostate is not used. More typically, prostatic epithelia is recombined with embryonic or neonatal mesenchyme from the urogenital area (either urogenital mesenchyme, UGM or seminal vesicle mesenchyme, SVM) and then implanted underneath the kidney capsule of nude or SCID mice. Pioneered by Cunha in the 1970s, this has proven to be a reliable and reproducible approach, although it has several limitations. Firstly, the site of xenografting is not within the prostate, but under the kidney capsule, because it is more easily accessible and has high vascularity. Secondly, the selected stroma is not matched for age, anatomical origin or even strain or species of the origin of epithelial cells; in fact some recombinants are heterospecific (being rat and mouse or mouse and human). Yet, the embryonic/neonatal stroma induces proliferation and instructs differentiation of prostatic epithelia and thus provides an effective experimental model to test the potential of putative stem cell populations. But is it the most appropriate niche to test stem cell potential? (Fig. 7.1).

Whilst human CD49f$^{hi}$/Trop2$^{hi}$ basal cells repopulated normal prostatic epithelium using this mouse-human recombinant approach with embryonic stroma (Goldstein et al. 2010), luminal cell fractions failed to produce viable epithelial structures. One interpretation of this result is that the luminal population does not have stem cells, but it is equally plausible that the stroma recombination bioassay was inadequate. This latter explanation is supported by the parallel study, using stem-cell enriched fractions of CARNs, where Wang et al. showed these rare luminal cells can give rise to prostatic epithelium using the mouse-mouse tissue recombination approach (Wang et al. 2009). Therefore, there is a need to be cautious when interpreting negative results and to be certain that failure to repopulate epithelial progeny is due to a lack of stem cell activity, rather than due to limitations of a bioassay.

Proof of the power of embryonic stroma lies in its ability to direct epithelial cell differentiation and even change lineage status as shown using human or mouse embryonic stem cells, to generate mature prostatic or bladder epithelium by tissue recombination (Taylor et al. 2006; Oottamasathien et al. 2007). These inductive and instructive stromal-stem cell signals are absent in adult stroma, or otherwise tissue homeostasis could not be achieved and tissue overgrowth would continue, underscoring the difference between embryonic and adult stroma signalling.

### *7.2.2 Assays to Identify Cells of Origin of Prostate Cancer*

With these caveats in mind, we now discuss the role of stroma in determining the cells of origin of prostate cancer. These cells are currently of great interest because

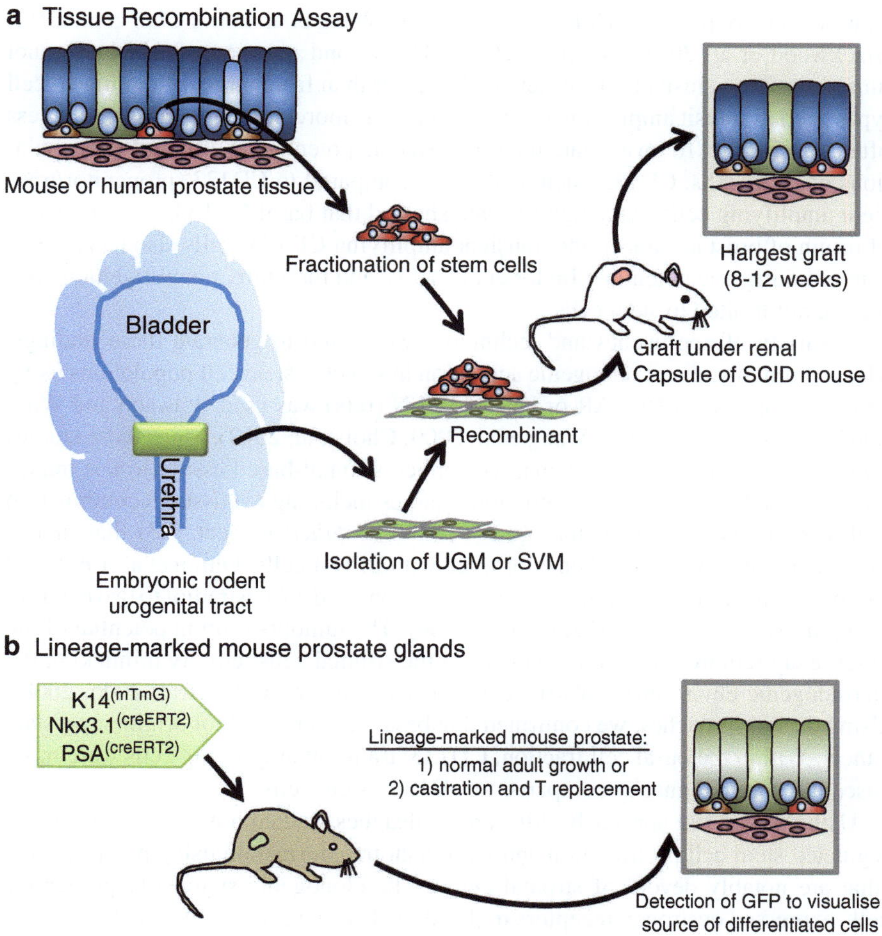


**Fig. 7.1** Different approaches used to identify prostatic epithelial stem cells and their progeny. (**a**) Tissue recombination utilises the inductive and instructive properties of rodent developmental (urogenital, UGM, or seminal vesicle, SVM) mesenchyme to induce differentiation in prospectively isolated sub-fractions of epithelium using cell surface markers (including CD133, CD44, Trop2, CD49f among others). (**b**) A complementary approach is genetic lineage marking/tracing in mice to determine the source of epithelial progeny using promoters such as PSA, cytokeratin 14 and/or Nkx3.1. This latter method relies on stromal and epithelial of the adult mouse prostate, which are significantly different to developmental mesenchyme, or castration and testosterone replacement to identify regenerating cells. Collectively, these approaches identified multiple stem cells in normal and tumour prostate tissues

they directly influence tumour phenotype and genotype. Two main questions in the field are as follows: (1) do both basal and luminal stem cells give rise to tumours and (2) are cells of origin restricted to stem cells in normal tissue, or are more differentiated (non-stem) cell types also susceptible to malignancy? Firstly, there is consistent evidence that basal and luminal stem cells from normal and non-malignant tissues can *both* act as cells of origin for prostate cancer if genetically modified

(Lawson and Witte 2007; Goldstein et al. 2010; Wang et al. 2009; Choi et al. 2012; Blackwood et al. 2011; Liu et al. 2011). The second question of whether or not tumours arise exclusively from stem cells (rather than from more differentiated cell types such as transit amplifying or luminal cells) is more difficult to answer and less often considered. To investigate the tumourigenic potential of different subpopulations of basal cells, $CD133^+$ stem cells were compared to $CD133^-$ progenitor/transient amplifying cells. Although the data showed that basal $CD133^+$ cells are a cell of origin of prostate cancer, the transient amplifying $CD133^-$ cells also had significant tumourigenic potential (Taylor et al. 2012), and therefore, tumourigenicity is a feature not limited to stem cells.

A variety of approaches and techniques were used to generate these findings. Most commonly, direct ongogenic activation in selected stem cell populations using activation of AKT, ERG, AR or loss of PTEN (refs) was used (Lawson and Witte 2007; Goldstein et al. 2010; Wang et al. 2009; Choi et al. 2012). Since these studies do not consider the role of stroma, we applied stromal-based assays to test malignant potential and complement previous studies including (a) tissue recombination with carcinoma-associated fibroblasts (CAFs) (*detailed in* Sect. 7.3) that induce malignant transformation when co-grafted with BPH-1 cells (Olumi et al. 1999) and (b) the administration of high doses of testosterone and 17β-estradiol to drive malignancy in tissue xenografts (Ricke et al. 2006). The tumour-forming potential of the latter assay requires signalling via ERα in the stromal cells, closely mimicking the steroidogenic environment observed in prostate cancer (Ricke et al. 2006, 2008). Using these approaches, we confirmed that basal cells are a cell of origin of prostate cancer. Within the basal cell fraction, $CD133^-$ transient amplifying cells were more susceptible to malignancy compared to $CD133^+$ stem cells.

Using a different approach, Prins and colleagues studied how steroid hormones regulated stem cell proliferation and malignant transformation using prostaspheres (that are notably devoid of stromal cells). The clonogenic stem/early progenitor cells expressed oestrogen receptors α, β and the G protein-coupled receptor 30, and in response to 17β-estradiol, the prostaspheres increased in number and size, showing them to be direct oestrogen cell targets (Hu et al. 2011b). Most importantly, in tissue recombination grafts human prostate progenitor cells were tumourigenic when stimulated with high-dose testosterone and 17β-estradiol, similar to our previous observations (Hu et al. 2011b; Taylor et al. 2012). Hormone action via stromal steroid receptors in stroma, rather than epithelia, remains to be fully elucidated, but it is clear that prostate stem/progenitor cells are highly susceptible to hormone-induced malignant transformation.

In combination, these data indicate that multiple cell types (i.e. basal and luminal/stem and progenitor) give rise to prostate cancer and show similarity to breast cancer, where multiple cell types give rise to distinct tumour types (Lim et al. 2009; Shackleton et al. 2006). In breast cancer, the differentiation hierarchy is better defined, and there are robust molecular profiles that can be correlated with the potential cells of origin, which are associated with a predicted outcome, and applicable therapeutic strategy (Visvader and Lindeman 2008). None of this type of information is available in prostate cancer, and the clinical relevance of the cellular origin of prostate cancer is not evident, because most tumours are adenocarcinomas and only rarely are neuroendocrine tumours. The main clinical need in prostate cancer is a predictor of tumour aggressiveness (Freedland 2011). A biological tool to predict tumour progression would significantly enhance the efficiency of our current treatments and limit overtreatment with complicated surgical interventions for indolent or nonaggressive disease. Whether the cellular origin of prostate cancer correlates to or is associated with clinical outcome is yet to be determined.

### *7.2.3 Role of Stromal in Identifying Prostate Cancer-Repopulating Cells*

The concept that minor tumour cell fractions with greater self-renewal and multi-lineage differentiation potential initiated debate in the cancer biology field (Reya et al. 2001). The suggestion that these cells were also resistant to chemotherapy and radiotherapy was an added reason to pursue their identity. Targeting of these cells would ensure effective long term cure from cancer by preventing any possible tumour regrowth or recurrence, as most therapies are directed at the fast growing tumour mass but not the slow-dividing cancer stem cells (Visvader and Lindeman 2012). It is plausible that these 'cancer-repopulating cells' are likely to be the root of tumour metastasis, even though they constitute a minority of the tumour itself, although this is unproven. Most recently, new evidence emerged that cells with metastatic potential undergo epithelial-to-mesenchymal transition and exhibit stem-like features as they prepare to exit the primary tumour site (Creighton et al. 2010; Zhang et al. 2012), supporting this concept of a major role for cancer-repopulating or stem cells in metastasis.

In prostate cancer, the identity of cancer stem (or repopulating) cells has not been fully elucidated. If we consider the three main properties of cancer stem cells, (1) self-renewal, (2) multi-lineage differentiation and (3) therapy resistance, some but not all of these parameters have been identified in various systems, but the presence of a single tumour cell that meets all three criteria has not been reported. Most of the work on prostate cancer-repopulating cells has been done using cell lines or xenograft lines that bear limited resemblance to the parental tumour. Common stem cells markers (i.e. CD133 and CD44) are easily detected in prostate cancer cell lines, LNCaP, DU145 and PC3, which typically display enriched tumour-forming potential compared to their negative fraction counterparts (Patrawala et al. 2007). These cells consistently show increased self-renewal and proliferative potential, although recent data suggests that not all drug-tolerant (or therapy-resistant) cells reside within these cell populations (Yan et al. 2011). Using primary specimens, prostate cancer-repopulating cells have been artificially generated using human prostate epithelial TERT, and their differentiation potential has been investigated in vitro and in vivo (Gu et al. 2007; Kasper 2007). The most relevant studies have been on prospectively isolated fresh cell fractions from primary prostate cancer specimens, showing that tumour cells with stem cell characteristics and repopulating potential are located in human tissues

(Collins et al. 2005; Toivanen et al. 2011); these tumour cells are highly dependent on surrounding stromal cues, unlike cell lines that have adapted to cell-autonomous growth. Subsequently, elegant lentivector tracing of human prostate cancer cells demonstrated populations of stem cells that robustly support tumour development and resist androgen ablation (Qin et al. 2012; Gaisa et al. 2011).

Maitland and colleagues demonstrated stem cell activity in primary specimens within the $CD44^{+}integrin\alpha2\beta1^{+}CD133^{+}$ fraction using assays for self-renewal and differentiation potential, although the transplantation assays in vivo have not yet been conducted (Collins et al. 2005), whilst Tang and colleagues showed $PSA^{lo}$-expressing cells to show repopulating activity (Qin et al. 2012). We recently enhanced the assay to test cancer stem cell activity using tissue recombination to enable dissociated cancer cells to be co-grafted with embryonic mouse mesenchyme, similar to that reported for normal stem cells (Toivanen et al. 2011). Using this assay, we tested the tumour repopulating potential of $\alpha2\beta1integrin^{hi}$ cells and showed that selection for this single cell surface marker did not influence the repopulating potential of primary cancer cells, confirming data obtained from xenograft cell lines (Patrawala et al. 2007). This assay, based on the inclusion of embryonic stroma, is now available to test the repopulating potential of isolated cell fractions of cancer cells derived from fresh primary specimens. However, as a note of caution, it is important to restate that this model relies on using stroma that is different to that which the cancer cells are exposed in vivo and further improvements using human stromal cells could advance the field.

Given there is lack of consensus on markers to identify cancer stem cells in human tissues, it has not been possible to fully determine the role of tumour stroma on directing their differentiation. This deficiency hampered studies to determine the importance of stromastem cell signalling in tumour tissue. In fact, the cell surface markers used to identify cancer cells with stem cell-like features have proven to be dynamic, and isolation using prospective sorting strategies will produce varying cell populations at any given time (Vander Griend et al. 2008).

Using the broad definition of cancer-repopulating cells being immune positive for AR, PSA and PSCA whilst not expressing the basal cell marker DNp63, Vander Griend and colleagues showed that AR promotes malignant growth of prostate cancer-repopulating cells via cell-autonomous signalling pathways, whereby AR acquires gain of function oncogenic ability to stimulate malignant growth (Vander Griend et al. 2010). This was postulated based on the finding that stromal AR expression was not required for prostate cancer growth, since tumour stroma surrounding AR-positive human prostate cancer metastases is characteristically AR negative (Wikstrom et al. 2009) and human AR-positive prostate cancer cells grew equally well when xenografted in wild-type vs. AR-null nude mice (Vander Griend et al. 2010). The mechanistic differences between stromal vs. cancer cell AR are poorly defined, and a greater understanding is essential to define the role of androgen actions in prostate cancer-repopulating or stem cells in hormone-naive disease.

One of the confounding issues with isolating cancer-repopulating cells from prostate cancer specimens is that unlike colon, breast or lung cancer, the pathological specimen is rarely all (100%) tumour. Prostate cancer is notoriously heterogeneous in its cellular composition, and tumour foci and benign glands sit adjacent to each other throughout the structure. The probability of obtaining a pure tumour sample is improbable (Priolo et al. 2010), and even with confirmation by examining frozen section at the time of removal, it is impossible to prove no benign tissue is present (Toivanen et al. 2011) and may compromise the quality of the data if this is not understood. At present, there is no single cell surface marker that can reliably distinguish benign from tumour epithelium, and so the cells derived from any primary specimen should be considered as mixed phenotypes. Likewise, the cell surface markers used to identify stem cells (especially $\alpha 2\beta 1$integrin$^{hi}$/CD44$^{+}$/CD133$^{+}$) are the same for normal and cancerous cells, so the situation is even more complex. This is also true for studies on tumour stroma that are derived from mixed pathologies, complicating studies on stroma-cancer stem cell interactions in human primary tissues that are required to advancing the field.

### *7.2.4 Stem Cells in Castration-Resistant Prostate Cancer*

Perhaps the most clinically relevant stem cells in prostate cancer are the castrate-resistant cells. These are the tumour cells that evade androgen deprivation therapy and are responsible for tumour regrowth and recurrence in those with advanced disease. Whilst there is speculation about the origins and/or evolution of these cells, there is little evidence to support any argument. These cells have simply not been identified in clinical specimens. A recent study using an established xenograft BM18 cell line showed that stem cell-like prostate cancer cells are selected by castration and survive as totally quiescent cells (Germann et al. 2012). Upon androgen replacement, the stem cell-like cells reinitiate BM18 tumour growth, confirming their repopulating potential. Since these were subcutaneous tumours, the contribution of stroma was not considered. It is unclear where castrate-resistant stem cells emerge under the pressure of androgen withdrawal, by an adaptation mechanism, or whether these repopulating cells are pre-existing and facilitate disease progression. Whatever is correct, the stroma should be considered as an active participant in their extrinsic regulation, particularly in response to androgen withdrawal, as it does in development. The paucity of knowledge about the genotype or phenotype of tumour stroma in castrate-resistant disease is an important area of need, as it will likely contribute significantly to novel therapeutic design for advanced disease.

## 7.3 Stromal Directed Differentiation and Malignant Transformation of Stem Cells

It is widely recognised that stroma-epithelial interactions underpin the differentiation and function of the prostate gland. As such, many groups intentionally use in vivo assays for stem cell studies in order to maintain these cell-cell interactions. However, as mentioned, there are technical limitations of the experimental stem cell assays used to assess prostate-regenerating potential. One important issue is that tissue recombination approaches fail to provide a naturally occurring stromal or niche environment to test stem cell regeneration potential. Most laboratories, ours included, use mismatched stroma, which represent an artificial situation and may stimulate stem cells differently, or inefficiently estimate the capability of stroma to elicit a stem cell response. It will be important to develop better models that more accurately mimic the stem cell-stromal interactions that occur in vivo, particularly in human, but this is currently a significant technical limitation.

Tumour stroma is part of the natural niche for cancer-repopulating cells. The normal stromal compartment has evolved with an inherent plasticity to respond rapidly to aberrant situations, including cancer, and act in concert with the adjacent epithelium leading to the emergence of 'reactive stroma'. Reactive stroma in prostate cancer is composed of CAFs and myofibroblasts which include remodelled matrix and altered expression of repair-associated growth factors and cytokines (Desmouliere et al. 2004; Gabbiani 2003). Prostatic CAFs can be isolated and cultured as primary cell lines, retaining their tumour promoting potential for a moderate but limited time span. These cells have been extensively characterised by multiple laboratories, in terms of phenotype and genotype, and are biologically distinct to their adjacent counterparts, normal prostatic fibroblasts obtained from the same patient specimen (Tuxhorn et al. 2002; Ayala et al. 2003; Ao et al. 2007; Joesting et al. 2005; Rowley and Barron 2012).

Tumour fibroblasts exhibit multiple similarities to developmental mesenchyme, reflecting a reawakening or reactivation of the growth regulatory systems and signalling pathways present in embryonic life, but quiescent in adulthood. However, tissue recombination studies using BPH-1 cells challenge this hypothesis; whilst human CAFs induce malignancy, UGM induces organised ductal structures of benign pathology (Wang et al. 2001; Taylor et al. 2012; Olumi et al. 1999). Defining the differences between developmental mesenchyme that stimulates proliferation and normal differentiation and CAFs that stimulate proliferation and carcinogenesis will reveal tumour-specific therapeutic targets.

It is important to consider that tumour stroma consists of more than CAFs, even though they are one of the predominant and functionally important cell types. It is widely accepted that progression of organ-confined tumours is influenced by angiogenesis and inflammatory cells, which contribute to the stem cell niche, providing a complex signalling environment. Interestingly, both the CAFs and prostate cancer stem cells mediate key regulatory pathways. For example, CAFs show elevated expression of key cytokines and chemokines, SDF-1, CXCL12 and CCL2, which contribute to an immune-rich microenvironment (Ao et al. 2007; Joesting et al. 2005). Likewise, prostate cancer stem cells are also highly responsive to immune modulation, and an immune signature was expressed in human $CD133^{+}$ cancer stem cells, including interleukin 6 (*IL6*) and interferon-γ receptor 1 (*IFGR1*) (Birnie et al. 2008). Studies defining the complex role of immune cell regulation of stem cells are required, but are reliant on new in vivo and in vitro approaches that allow these multicell interactions to be maintained.

Whilst the major focus of this chapter has been on human models, mouse models of prostate cancer where only stroma is genetically modified provide convincing proof of the biological importance of stroma. Single or multiple gene mutations in stromal fibroblasts result in abnormal pathologies in the adjacent glandular epithelium of the mice. This was demonstrated with overexpression of FGF10 in a tissue recombination model (Memarzadeh et al. 2007) and TGFβRII in transgenic mice (Bhowmick et al. 2004). The latter has developed into a reliable model of prostate intra-epithelia neoplasia which progresses to adenocarcinoma (Cheng et al. 2005; Bhowmick et al. 2004). This severe pathology resulted from the deletion of a single growth factor signalling pathway, in stromal cells alone. The finding that tumours were resistant to castration broadens the biological implications of this model and provides a system to study the effects of stromal signalling on stem cell differentiation and their role in malignant transformation and castrate-resistant disease.

## 7.4 Summary/Conclusions/Future Directions

The purpose of this review has been to highlight cells with significant therapeutic potential, as they hold the key to improved clinical treatments for men with prostate cancer. We propose that the regulation of stem cells in prostate cancer is a combination of intrinsic or extrinsic signalling. Further understanding of the cellular biology and mechanistic approaches to arrest tumour growth will be essential to improving clinical outcomes for patients. We have presented a case to suggest both stem cells in cancer (either cells of origin or cancer stem cells) and their surrounding stromal cells are active contributors to tumourigenesis and are thus valid therapeutic targets. This is not unique to prostate cancer, but is typical of many solid tumours, including hormone-dependent cancers of the breast, endometrium and ovary.

We propose that the way forward will include better model systems, which allow alignment of stroma and stem cells. Whilst this is a simple concept, it is technically challenging since the identity of human stem cells is not clearly defined. Additionally, whilst the effects of CAFs are significant, there are other stromal components that comprise the stem cell niche, which should also be examined. Doing this in vivo is even more complicated because the host endogenous stromal contribution also contributes to the phenotype.

It is probable that the development of improved in vitro coculture or 3D model systems to directly test the stromal-stem cell interactions will be as useful and informative. Such approaches are under development in collaborations between biologists and bioengineers that generate smart bio-scaffolds to replicate the in vivo situation (Hutmacher 2010; Sieh et al. 2010). These base models can then be built upon to include the individual tumour components including subtypes of immune cells, endothelial cells or lymphatics or vasculature that contribute to stem cell behaviour in cancer.

## References

Ao M, Franco OE, Park D, Raman D, Williams K, Hayward SW (2007) Cross-talk between paracrine-acting cytokine and chemokine pathways promotes malignancy in benign human prostatic epithelium. Cancer Res 67:4244–4253

Ayala G, Tuxhorn JA, Wheeler TM, Frolov A, Scardino PT, Ohori M, Wheeler M, Spitler J, Rowley DR (2003) Reactive stroma as a predictor of biochemical-free recurrence in prostate cancer. Clin Cancer Res 9:4792–4801

Bhowmick NA, Chytil A, Plieth D, Gorska AE, Dumont N, Shappell S, Washington MK, Neilson EG, Moses HL (2004) TGF-beta signaling in fibroblasts modulates the oncogenic potential of adjacent epithelia. Science 303:848–851

Birnie R, Bryce SD, Roome C, Dussupt V, Droop A, Lang SH, Berry PA, Hyde CF, Lewis JL, Stower MJ, Maitland NJ, Collins AT (2008) Gene expression profiling of human prostate cancer stem cells reveals a pro-inflammatory phenotype and the importance of extracellular matrix interactions. Genome Biol 9:R83

Blackwood JK, Williamson SC, Greaves LC, Wilson L, Rigas AC, Sandher R, Pickard RS, Robson CN, Turnbull DM, Taylor RW, Heer R (2011) In situ lineage tracking of human prostatic epithelial stem cell fate reveals a common dual origin for basal and luminal cells. J Pathol 225:181–188

Burger PE, Xiong X, Coetzee S, Salm SN, Moscatelli D, Goto K, Wilson EL (2005) Sca-1 expression identifies stem cells in the proximal region of prostatic ducts with high capacity to reconstitute prostatic tissue. Proc Natl Acad Sci USA 102:7180–7185

Cheng N, Bhowmick NA, Chytil A, Gorksa AE, Brown KA, Muraoka R, Arteaga CL, Neilson EG, Hayward SW, Moses HL (2005) Loss of TGF-beta type II receptor in fibroblasts promotes mammary carcinoma growth and invasion through upregulation of TGF-alpha-, MSP- and HGF-mediated signaling networks. Oncogene 24:5053–5068

Choi N, Zhang B, Zhang L, Ittmann M, Xin L (2012) Adult murine prostate basal and luminal cells are self-sustained lineages that can both serve as targets for prostate cancer initiation. Cancer Cell 21:253–265

Collins AT, Berry PA, Hyde C, Stower MJ, Maitland NJ (2005) Prospective identification of tumorigenic prostate cancer stem cells. Cancer Res 65:10946–10951

Creighton CJ, Chang JC, Rosen JM (2010) Epithelial-mesenchymal transition (EMT) in tumor-initiating cells and its clinical implications in breast cancer. J Mammary Gland Biol Neoplasia 15:253–260

Desmouliere A, Chaponnier C, Gabbiani G (2004) The stroma reaction myofibroblast: a key player in the control of tumor cell behaviour. Int J Dev Biol 48:509–517

English HF, Santen RJ, Isaacs JT (1987) Response of glandular versus basal rat ventral prostatic epithelial cells to androgen withdrawal and replacement. Prostate 11:229–242

Freedland SJ (2011) Screening, risk assessment, and the approach to therapy in patients with prostate cancer. Cancer 117:1123–1135

Gabbiani G (2003) The myofibroblast in wound healing and fibrocontractive diseases. J Pathol 4:500–503

Gaisa NT, Graham TA, McDonals SA, Poulsom R, Heidenreich A, Jakse G, Knuechel R, Wright NA (2011) Clonal architechture of human prostatic epithelium in benign and malignant conditions. J Pathol 225:172–180

Germann M, Wetterwald A, Guzman-Ramirez N, van der Pluijm G, Culig Z, Cecchini MG, Williams ED, Thalmann GN (2012) Stem-like cells with luminal progenitor phenotype survive castration in human prostate cancer. Stem Cells 30:1076–1086

Goldstein AS, Huang J, Guo C, Garraway IP, Witte ON (2010) Identification of a cell of origin for human prostate cancer. Science 329:568–571

Goldstein AS, Lawson DA, Cheng D, Sun W, Garraway IP, Witte ON (2008) Trop2 identifies a subpopulation of murine and human prostate basal cells with stem cell characteristics. Proc Natl Acad Sci USA 105:20882–20887

Goodyear SM, Amatangelo MD, Stearns ME (2009) Dysplasia of human prostate CD133(hi) subpopulation in NOD-SCIDS is blocked by c-myc anti-sense. Prostate 69(7):689–698
Gu G, Yuan J, Wills M, Kasper S (2007) Prostate cancer cells with stem cell characteristics reconstitute the original human tumor in vivo. Cancer Res 67:4807–4815
Hu WY, Shi GB, Hu DP, Nelles JL, Prins GS (2011a) Actions of estrogens and endocrine disrupting chemicals on human prostate stem/progenitor cells and prostate cancer risk. Mol Cell Endocrinol 354(1–2):63–73
Hu WY, Shi GB, Lam HM, Hu DP, Ho SM, Madueke IC, Kajdacsy-Balla A, Prins GS (2011b) Estrogen-initiated transformation of prostate epithelium derived from normal human prostate stem-progenitor cells. Endocrinology 152:2150–2163
Hutmacher DW (2010) Biomaterials offer cancer research the third dimension. Nat Mater 9:90–93
Joesting MS, Perrin S, Elenbaas B, Fawell SE, Rubin JS, Franco OE, Hayward SW, Cunha GR, Marker PC (2005) Identification of SFRP1 as a candidate mediator of stromal-to-epithelial signaling in prostate cancer. Cancer Res 65:10423–10430
Kasper S (2007) Characterizing the prostate stem cell. J Urol 178:375
Lawson DA, Witte ON (2007) Stem cells in prostate cancer initiation and progression. J Clin Invest 117:2044–2050
Lawson DA, Xin L, Lukacs RU, Cheng D, Witte ON (2007) Isolation and functional characterization of murine prostate stem cells. Proc Natl Acad Sci USA 104:181–186
Leong KG, Wang BE, Johnson L, Gao WQ (2008) Generation of a prostate from a single adult stem cell. Nature 456:804–808
Lim E, Vaillant F, Wu D, Forrest NC, Pal B, Hart AH, Asselin-Labat ML, Gyorki DE, Ward T, Partanen A, Feleppa F, Huschtscha LI, Thorne HJ, Fox SB, Yan M, French JD, Brown MA, Smyth GK, Visvader JE, Lindeman GJ (2009) Aberrant luminal progenitors as the candidate target population for basal tumor development in BRCA1 mutation carriers. Nat Med 15:907–913
Liu J, Pascal LE, Isharwal S, Metzger D, Ramos Garcia R, Pilch J, Kasper S, Williams K, Basse PH, Nelson JB, Chambon P, Wang Z (2011) Regenerated luminal epithelial cells are derived from preexisting luminal epithelial cells in adult mouse prostate. Mol Endocrinol 25:1849–1857
Lukacs RU, Goldstein AS, Lawson DA, Cheng D, Witte ON (2010) Isolation, cultivation and characterization of adult murine prostate stem cells. Nat Protoc 5:702–713
Memarzadeh S, Xin L, Mulholland DJ, Mansukhani A, Wu H, Teitell MA, Witte ON (2007) Enhanced paracrine FGF10 expression promotes formation of multifocal prostate adenocarcinoma and an increase in epithelial androgen receptor. Cancer Cell 12:572–585
Miki J, Furusato B, Li H, Gu Y, Takahashi H, Egawa S, Sesterhenn IA, McLeod DG, Srivastava S, Rhim JS (2007) Identification of putative stem cell markers, CD133 and CXCR4, in hTERT-immortalized primary nonmalignant and malignant tumor-derived human prostate epithelial cell lines and in prostate cancer specimens. Cancer Res 67:3153–3161
Olumi AF, Grossfeld GD, Hayward SW, Carroll PR, Tlsty TD, Cunha GR (1999) Carcinoma-associated fibroblasts direct tumor progression of initiated human prostatic epithelium. Cancer Res 59:5002–5011
Oottamasathien S, Wang Y, Williams K, Franco OE, Wills ML, Thomas JC, Saba K, Sharif-Afshar AR, Makari JH, Bhowmick NA, Demarco RT, Hipkens S, Magnuson M, Brock JW III, Hayward SW, Pope JCT, Matusik RJ (2007) Directed differentiation of embryonic stem cells into bladder tissue. Dev Biol 304:556–566
Patrawala L, Calhoun-Davis T, Schneider-Broussard R, Tang DG (2007) Hierarchical organization of prostate cancer cells in xenograft tumors: the CD44+alpha2beta1+ cell population is enriched in tumor-initiating cells. Cancer Res 67:6796–6805
Priolo C, Agostini M, Vena N, Ligon AH, Fiorentino M, Shin E, Farsetti A, Pontecorvi A, Sicinska E, Loda M (2010) Establishment and genomic characterization of mouse xenografts of human primary prostate tumors. Am J Pathol 176(4):1901–1913

Qin J, Liu X, Laffin B, Chen X, Choy G, Jeter CR, Calhoun-Davis T, Li H, Palapattu GS, Pang S, Lin K, Huang J, Ivanov I, Li W, Suraneni MV, Tang D (2012) The PSA-/lo prostate cancer cell population harbors self-renewing long-term tumor-propagating cells that resist castration. Cell Stem Cell 10:556–569

Reya T, Morrison SJ, Clarke MF, Weissman IL (2001) Stem cells, cancer, and cancer stem cells. Nature 414:105–111

Richardson GD, Robson CN, Lang SH, Neal DE, Maitland NJ, Collins AT (2004) CD133, a novel marker for human prostatic epithelial stem cells. J Cell Sci 117:3539–3545

Ricke WA, Ishii K, Ricke EA, Simko J, Wang Y, Hayward SW, Cunha GR (2006) Steroid hormones stimulate human prostate cancer progression and metastasis. Int J Cancer 118:2123–2131

Ricke WA, McPherson SJ, Bianco JJ, Cunha GR, Wang Y, Risbridger GP (2008) Prostatic hormonal carcinogenesis is mediated by in situ estrogen production and estrogen receptor alpha signaling. FASEB J 22:1512–1520

Rowley D, Barron DA (2012) The reactive stroma microenvironment and prostate cancer progression. Endocr Relat Cancer 19(6):R187–R204

Shackleton M, Vaillant F, Simpson KJ, Stingl J, Smyth GK, Asselin-Labat ML, Wu L, Lindeman GJ, Visvader JE (2006) Generation of a functional mammary gland from a single stem cell. Nature 439:84–88

Sieh S, Lubik AA, Clements JA, Nelson CC, Hutmacher DW (2010) Interactions between human osteoblasts and prostate cancer cells in a novel 3D in vitro model. Organogenesis 6:181–188

Taylor RA, Cowin PA, Cunha GR, Pera M, Trounson AO, Pedersen J, Risbridger GP (2006) Formation of human prostate tissue from embryonic stem cells. Nat Methods 3:179–181

Taylor RA, Toivanen R, Frydenberg M, Pedersen J, Harewood L, Australian Prostate Cancer Biosource, Collins AT, Maitland NJ, Risbridger GP (2012) Human epithelial basal cells are cells of origin of prostate cancer, independent of CD133 status. Stem Cells 30(6):1087–1096

Taylor RA, Toivanen R, Risbridger GP (2010) Stem cells in prostate cancer: treating the root of the problem. Endocr Relat Cancer 17:R273–R285

Toivanen R, Berman DM, Wang H, Pedersen J, Frydenberg M, Meeker AK, Ellem SJ, Risbridger GP, Taylor RA (2011) Brief Report—a bioassay to Identify primary human prostate cancer repopulating cells. Stem Cells 29(8):1310–1314

Tuxhorn JA, Ayala GE, Smith MJ, Smith VC, Dang TD, Rowley DR (2002) Reactive stroma in human prostate cancer: induction of myofibroblast phenotype and extracellular matrix remodeling. Clin Cancer Res 8:2912–2923

Vander Griend DJ, D'Antonio J, Gurel B, Antony L, Demarzo AM, Isaacs JT (2010) Cell-autonomous intracellular androgen receptor signaling drives the growth of human prostate cancer initiating cells. Prostate 70:90–99

Vander Griend DJ, Karthaus WL, Dalrymple S, Meeker A, Demarzo AM, Isaacs JT (2008) The role of CD133 in normal human prostate stem cells and malignant cancer-initiating cells. Cancer Res 68:9703–9711

Visvader JE, Lindeman GJ (2008) Cancer stem cells in solid tumours: accumulating evidence and unresolved questions. Nat Rev Cancer 8:755–768

Visvader JE, Lindeman GJ (2012) Cancer stem cells: current status and evolving complexities. Cell Stem Cell 10:717–728

Wang X, Kruithof-De Julio M, Economides KD, Walker D, Yu H, Halili MV, Hu YP, Price SM, Abate-Shen C, Shen MM (2009) A luminal epithelial stem cell that is a cell of origin for prostate cancer. Nature 461:495–500

Wang Y, Sudilovsky D, Zhang B, Haughney PC, Rosen MA, Wu DS, Cunha TJ, Dahiya R, Cunha GR, Hayward SW (2001) A human prostatic epithelial model of hormonal carcinogenesis. Cancer Res 61:6064–6072

Wikstrom P, Marusic J, Stattin P, Bergh A (2009) Low stroma androgen receptor level in normal and tumor prostate tissue is related to poor outcome in prostate cancer patients. Prostate 69:799–809

Xin L, Ide H, Kim Y, Dubey P, Witte ON (2003) In vivo regeneration of murine prostate from dissociated cell populations of postnatal epithelia and urogenital sinus mesenchyme. Proc Natl Acad Sci USA 100:11896–11903

Xin L, Lawson DA, Witte ON (2005) The Sca-1 cell surface marker enriches for a prostate-regenerating cell subpopulation that can initiate prostate tumorigenesis. Proc Natl Acad Sci USA 102:6942–6947

Xin L, Lukacs RU, Lawson DA, Cheng D, Witte ON (2007) Self-renewal and multilineage differentiation in vitro from murine prostate stem cells. Stem Cells 25:2760–2769

Yan H, Chen X, Zhang Q, Qin J, Li H, Liu C, Calhoun-Davis T, Coletta LD, Klostergaard J, Fokt I, Skora S, Priebe W, Bi Y, Tang DG (2011) Drug-tolerant cancer cells show reduced tumor-initiating capacity: depletion of CD44 cells and evidence for epigenetic mechanisms. PLoS One 6:e24397

Zhang Y, Wei J, Wang H, Xue X, An Y, Tang D, Yuan Z, Wang F, Wu J, Zhang J, Miao Y (2012) Epithelial mesenchymal transition correlates with CD24+CD44+ and CD133+ cells in pancreatic cancer. Oncol Rep 27:1599–1605

# Chapter 8
# Targeting the Prostate Stem Cell for Chemoprevention

**Molishree U. Joshi, Courtney K. von Bergen, and Scott D. Cramer**

**Abstract** Prostate cancer is the most common cancer diagnosis and the second deadliest non-cutaneous cancer in men in the western world. Despite the advancement of new therapies, survival rates remain tightly correlated to the stage in which a patient is diagnosed. The heterogeneous nature of tumor and their variable response to therapies is now being attributed to tumor-initiating cells or cancer stem cells. The cancer stem cell hypothesis suggests that cancer initiates from a specific subset of tumor cells that possess "stemlike" properties. Over the past several decades, the use of natural or synthetic agents such as vitamins, foods, or spices has been shown to correlate with lower incidences of cancer. Many of these chemopreventive agents have now been reported to cause differentiation of cancer stem cells and suppress their proliferation, thereby making them more amenable to conventional therapies. This chapter will discuss the use of chemoprevention to target cancer stem cells and how these approaches can be applied to prostate cancer.

## 8.1 Introduction

Despite the advances in screening and early diagnosis, prostate cancer (PCa) remains a leading cause of cancer-related deaths in the western world. One in six men in the USA will be diagnosed with prostate cancer in his lifetime, and one in eight diagnosed will die from this disease. In 2010 alone, over 33,000 men died of prostate cancer in the USA (Siegel 2012). Early diagnosis can often lead to disease-free survival; however, survival rates decrease drastically in advanced disease. In addition, even survivors of early stage prostate cancer face a lifetime of

M.U. Joshi, Ph.D. • C.K. von Bergen, Ph.D. • S.D. Cramer, Ph.D. (✉)
Department of Pharmacology, University of Colorado, Anschutz Medical Campus, Aurora, CO, USA
e-mail: scott.cramer@ucdenver.edu

S.D. Cramer (ed.), *Stem Cells and Prostate Cancer*, DOI 10.1007/978-1-4614-6498-3_8, 

complications from treatment such as urinary incontinence and erectile dysfunction. As PCa is a slow-growing tumor that affects men late in life, delaying its initiation and progression by 5–10 years, by use of chemoprevention, could save thousands of lives each year.

The factors leading to prostate cancer development and progression are poorly understood; cellular mechanisms that drive prostate cancer initiation and progression continue to be investigated in an effort to find new ways of better fighting this disease. Tumor-targeted therapies have been met with many challenges, and the existence of cancer stem cells (CSCs) in tumors offers an attractive reason for failed therapies. The CSC hypothesis (reviewed in Ito et al. 2012; Li and Tang 2011) suggests that cancer initiates from a mutated stem cell in the tissue or from dedifferentiation of mature (cancer) cell. Like normal stem cells, CSC or tumor-initiating cells (TICs) have the potential to self-renew and differentiate, giving rise to tumors (Dick 2008; Donnenberg and Donnenberg 2005; Ito et al. 2012; Maitland and Collins 2008; Visvader and Lindeman 2008). Additionally, they possess innate resistance to chemotherapeutic and radiation treatment and comprise a limited targetable population of TICs in tumors.

Epidemiological studies have suggested age, race, family history, and hormone status as risk factors for developing PCa (Strope and Andriole 2010). There is also a correlation between diet/dietary supplementation and prostate cancer risk (Klein 2005; Syed et al. 2008; Venkateswaran and Klotz 2010). Various natural and synthetic compounds have demonstrated the ability to induce differentiation of CSCs, thereby making them more targetable. Due to these issues, as well as an interest in developing less toxic and more effective therapies, cancer research has turned to chemoprevention with reasonable success.

## 8.2 Cancer Stem Cell Hypothesis

Tumors are comprised of functionally and phenotypically heterogeneous cells. At least two models have been put forward to account for the manifestation of these differences. Although a detailed description of these models is beyond the scope of this chapter, they are widely reviewed (Li and Tang 2011; Subramaniam et al. 2010; Tu and Lin 2012; Visvader and Lindeman 2008), discussed in Chap. 1, and summarized in Fig. 8.1. A normal cellular hierarchy comprises of stem cell (at the apex), which has the potential to self-renew as well as progressively generate common and restricted progenitors, and mature cells (Fig. 8.1a). In contrast, tumor heterogeneity is attributed to either (1) clonal evolution (stochastic model) or (2) CSC (hierarchical model). According to the clonal evolution model, tumors are biologically homogeneous, and all undifferentiated cells have similar tumorigenic capacity. Any functional variations are thought to arise due to intrinsic (genetic, epigenetic) and extrinsic (environmental) influences (Fig. 8.1b). In contrast, the CSC model proposes hierarchical organization of cells in tumors (similar to normal tissue), in which a subpopulation of cells has the ability to initiate tumor growth. These CSC possess the ability to self-renew and give rise to non-tumorigenic progeny that

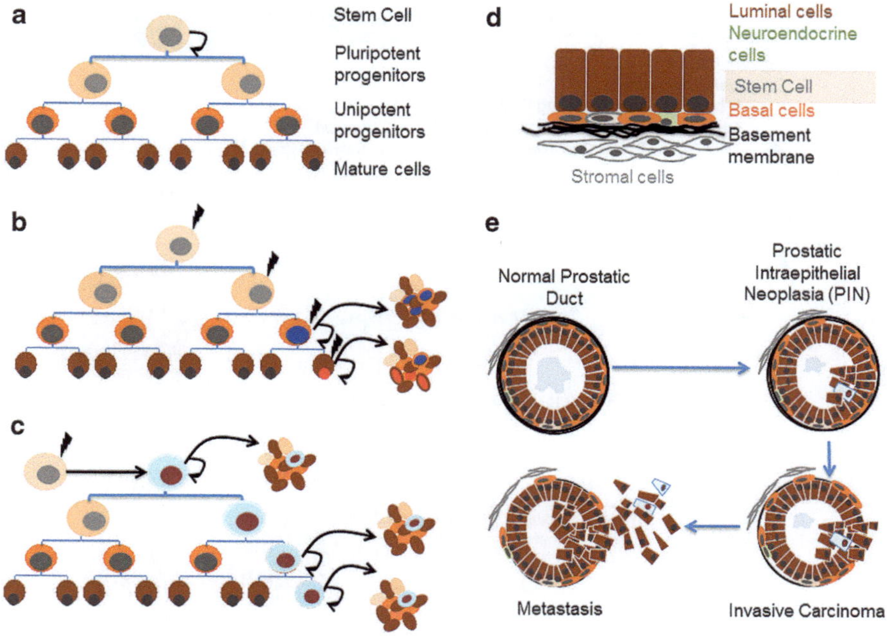


**Fig. 8.1** Models for normal and cancer stem cell propagation and its effect on tissue architecture. (**a**) Normal stem cells (at apex) maintain their population by self-renewal and also replenish mature cell population by generating progenitors with variable potency. In prostate, the transit amplifying cells represent the pluripotent subset. (**b**) Clonal evolution or stochastic model proposes that all undifferentiated cells in the tumor have the potential to initiate tumor growth. Additionally it is thought that mature cells could acquire mutations that revive their self-renewal capabilities and hence endow on them tumorigenic capacity. (**c**) Cancer stem cell hypothesizes that only a subset of cells, i.e., CSC, can generate tumors. (**d**, **e**) Schematic representation of stem and mature cells in prostate ducts. (**e**) Hyperplasia of epithelial cells in prostatic duct progressively leads to PIN, invasive carcinoma, and eventually metastasis. This process usually takes place over a long period and offers an excellent opportunity for chemoprevention (figure adapted from Abate-Shen and Shen 2000; Visvader and Lindeman 2008)

make up the bulk of tumor (Fig. 8.1c and Chap. 1). When normal stem cells mutate into a CSC, they continue to possess the inherent properties to self-renew and differentiate. Whereas if mature (or progenitor) cells transform into a CSC, they most likely acquire "stemlike" properties. It is possible that both models fuel tumor maintenance. Irrespective of the mechanism that gives rise to CSC, they possess genetic and epigenetic alterations leading to modifications in cell-surface and metabolic markers, gene expression, signaling pathways, and cellular fate (proliferation, senescence, apoptosis, autophagy, etc.) (Fig. 8.2). In prostate tissue, differentiation of normal stem cells leads to appropriate organization of mature cells (Fig. 8.1d) into functional ductal structure (Fig. 8.1e). This tissue organization is lost over time due to aberrant cell proliferation (Fig. 8.1e). In light of the current evidence, it is reasonable to postulate that this aberrant cellular behavior is most likely initiated and maintained by CSCs.

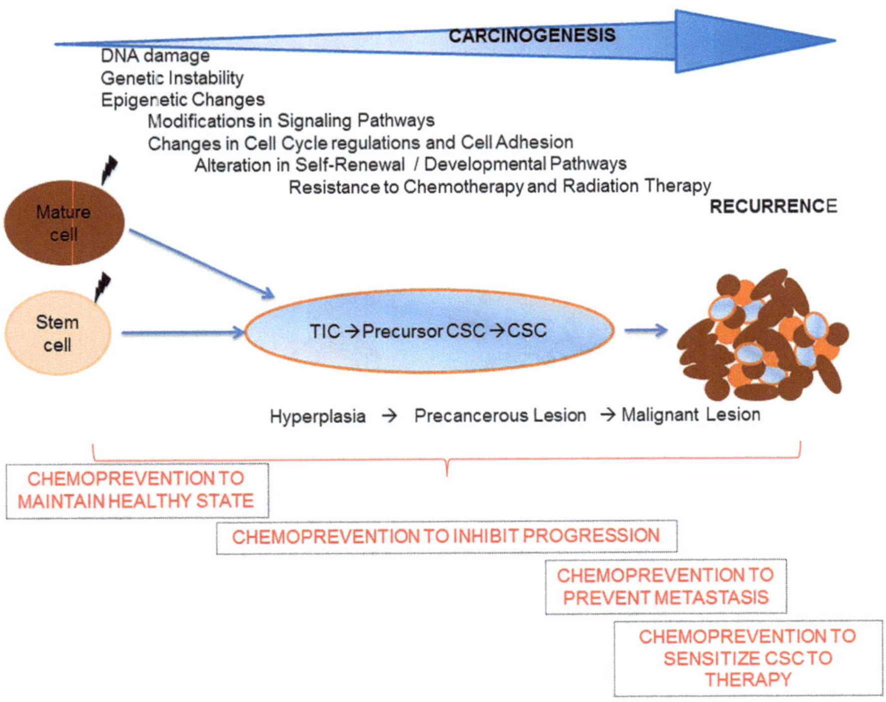


**Fig. 8.2** Schematic diagram summarizes the cellular processes involved in transformation of a normal stem cell to a CSC via stages of tumor-initialing cell (TIC) and precursor CSC. These cellular stages are most likely associated with the pathological cancer stages. Chemopreventive intervention has demonstrated the potential to inhibit carcinogenesis by altering key pathway that prevents self-renewal of CSCs and stimulates their differentiation, thereby making them sensitive to conventional therapy

## 8.3 Chemoprevention Targets for Cancer Stem Cells

The study of carcinogenesis has led to the current dogma that human carcinogenesis is a multiyear process involving multiple hits. This slow process provides the opportunity to intervene prior to accumulation of mutations and/or phenotypic changes, and chemoprevention can be the tool to accomplish this. The term "chemoprevention" was first coined in 1976 by Sporn et al. (1976) to define "the specific use of agents to reverse, suppress or prevent the carcinogenic process to invasive cancer" (Fig. 8.2). Thus, chemoprevention is a means to slow the process of carcinogenesis by the use of natural or synthetic agents such as vitamins, foods, or spices. Many ancient cultures believed that certain natural substances contained medicinal properties. Current scientific research has been focused on understanding how these compounds function and how these might be used in modern medicine. It is the hope that by targeting prostate cancer-associated stem cells or TICs with chemopreventive methods, fewer cases of prostate cancer will be clinically diagnosed and tumors will become more treatable or less aggressive. CSC-targeted chemoprevention studies and their overall findings are summarized in Table 8.1. Significant advances have been made in targeting stem cells in many systems such as breast, colon, blood, liver, brain, and lungs (Izrailit and Reedijk 2012; Kavalerchik et al. 2008; Ricci-Vitiani et al. 2009; Yi and Nan 2008); however, studies in the prostate have lagged behind. Applying concepts discovered in other systems to prostate cancer may assist in advancing this field. Additionally, no clinical trials for any cancer have shown efficacy for chemopreventive agents. This is likely due to multiple factors, including inappropriate design and lack of statistical power.

### *8.3.1 Preventing Cancer Initiation by Maintaining Normal Stemness*

Stem cells survive in a tissue for much longer than typical differentiated cells. During this time, stem cells must respond to environmental and physical challenges in order to maintain homeostasis of a tissue. Any alterations to their tightly regulated signaling made by mutations or epigenetic modifications could potentially stimulate tumor initiation. The ability of a chemopreventive agent to obstruct transformation of normal stem cells could theoretically prevent the beginnings of tumor formation.

Chemopreventive agents may help repair DNA damage and assist in maintaining cellular genomic integrity when challenged by mutations and epigenetic modifications. Resveratrol, found in red grapes and red wine, prevented DNA damage from radiation in mouse embryonic stem cells. In these cells, repair was accelerated, generation of reactive oxygen species was minimal, and genomic integrity was not compromised (Denissova et al. 2012). Resveratrol also promoted survival of primary endothelial stem cells in culture. In a dose-dependent manner, these cells were able to evade culture-induced senescence by maintaining their stem cell characteristics, possibly from increased telomerase activity (Wang et al. 2011). These data demonstrate that a naturally occurring chemopreventive agent has the potential to impede transformation of NSC into TIC or CSC, thereby preventing initiation of tumorigenesis.

### *8.3.2 Preventing Cancer Progression by Targeting Cancer Stem Cells*

CSCs are involved in all stages of cancer development, and chemopreventive agents have been shown to inhibit progression of already developing lesions by targeting the self-renewing capacity of CSC and by stimulating their differentiation. CSCs have the ability to self-propagate, and chemoprevention has exhibited the ability to

**Table 8.1** Chemopreventive agents and their response in cancer stem cells

| Chemopreventive agent | Food origin | Cancer/CSC | Target/mechanism of action | Reference |
|---|---|---|---|---|
| Auraptene | Fruits and edible plants | Colon ($CD116^+$ $CD44^+$ HT-29 and HCT-116 cell lines) | Down regulation of EGFR | Epifano et al. (2012) |
| Curcumin | Turmeric | Breast ($ALDH^+$ MCF7 and SUM159 cell lines) | Inhibition of Wnt signaling | Kakarala et al. (2010) |
| | | Leukemia | Inhibition of drug resistance transporter | Anuchapreeda et al. (2006) |
| | | Glioma (SP C6 cell line) | | Fong et al. (2010) |
| | | Brain ($CD133^+$ DAOY cell line) | Down regulation of STAT3 or IGF | Lim et al. (2011) |
| | | Colon ($CD116^+$ $CD44^+$ HT-29 and HCT-116 cell lines) | Inhibition of STAT3 | Yu et al. (2009) |
| | | Ovaries (KB-V-1) and Breast (MCF7) | Inhibition of drug resistance transporter | Limtrakul et al. (2007) |
| Epigallochtechin-3-gallate (EGCG) | Green tea | Pancreas ($CD133^+$ $CD44^+$ $CD24^+$ $ESA^+$ cells) | Apoptosis, inhibition of sonic hedgehog, inhibition of EMT | Tang et al. (2012) |
| | | Endothelium | Down regulation of MMP-9 | Ohga et al. (2009) |
| | | Prostate ($CD44^+CD133^+\alpha2\beta1^+$ cells, PC3 and LNCaP cell lines) | Apoptosis, inhibition of EMT | Tang et al. (2010) |
| | | Breast ($CD44^+$ MDA-MB-231cell line) | Activation of AMPK | Chen et al. (2012) |
| Genistein | Soy | Prostate ($CD44^+$ DU145 and 22RV1 cell lines) | Inhibition of hedgehog/Gli | Zhang et al. (2012) |
| | | Prostate (LNCaP) | Inhibition of EMT | Zhang et al. (2008) |
| | | Breast (MCF7, MDA-MB-231) | | Montales et al. (2012) |
| Piperine | Black pepper | Breast ($ALDH^+$ MCF7 and SUM159) | Inhibition of Wnt signaling | Kakarala et al. (2010) |
| PSP | Turkey tail | Prostate ($CD133^+$ $CD44^+$ PC3 cell line) | Inhibition of in vivo tumor growth | Luk et al. (2011) |

| | | | | |
|---|---|---|---|---|
| Quercetin | Plant flavonoid | Pancreas (ALDH+) | Apoptosis, inhibition of EMT, inhibition of in vivo tumor growth | Zhou et al. (2010) |
| Resveratrol | Grapes, berries, plum | Pancreas ($CD133^+$ $CD44^+$ $CD24^+$ $ESA^+$ human pancreatic cells) | Inhibition of pluripotency genes, inhibition of drug resistance gene, inhibition of EMT | Shankar et al. (2011) |
| | | Breast ($CD24^-CD44^+$ $ESA^+$ cells) | Apoptosis, inhibition of lipid synthesis | Pandey et al. (2011) |
| | | Bone (RAW 264.7) | Apoptosis | He et al. (2010) |
| | | Endothelium (EPC) | | Wang et al. (2011) |
| Sulforaphane | Cruciferous vegetable | Breast ($ALDH^+$ MCF7 and SUM159 cell lines) | Down regulation of Wnt/β-catenin | Li et al. (2010b) |
| | | Pancreases | Down regulation of NFκB | Kallifatidis et al. (2011) |
| | | Blood | Apoptosis | Fimognari et al. (2008) |
| | | Endothelium | Apoptosis | Nishikawa et al. (2009) |
| | | Prostate (DU145) | Promotes drug sensitivity | Kallifatidis et al. (2011) |
| γ-Tocotrienol (T3) | Vitamin E | Prostate ($CD133^+$ $CD44^+$ PC3 and DU145 cell lines) | Apoptosis | Luk et al. (2011); Zhou et al. (2010) |
| Vitamin D3 | Fish, eggs, etc. | Prostate (mouse prostate progenitor) | Inhibition of proliferation, cell cycle arrest, differentiation | Maund et al. (2011) |
| | | Bone | Differentiation | D'Ippolito et al. (2002); Piek et al. (2010) |
| | | Breast ($CD44^+$ MCF10DCIS) | Apoptosis, inhibition of tumor growth in vivo | So et al. (2011) |
| NSAID (Sulindac) | – | Intestine (Lgr5+ *APCMin/*$^+$crypt cells) | Apoptosis, down regulation of Wnt/β-catenin | Qiu et al. (2010) |

apply the brakes on this process by enhancing multi-lineage differentiation. To be able to better target or enhance these cellular processes, it is important to identify the molecular signaling responsible for these pathways. The Wnt, Hedgehog, Notch, and transforming growth factor-beta (TGF-β) pathways have been implicated in the process of self-renewal (Maund and Cramer 2010). Chemoprevention can alter the expression of stem cell markers, expressed by TICs or CSCs, used as a measure of cellular differentiation. Although no universal CSC markers have been defined, several cell-surface markers such as CD44, CD133, CD116, aldehyde dehydrogenases (ALDH), and Lrg5 have been reported to be overexpressed in various solid organ tumors and used to identify CSCs (Subramaniam et al. 2010).

#### 8.3.2.1 Targeting Self-Renewal

Chemoprevention has demonstrated the ability to directly block tumor growth in vivo. Chemopreventive agents aim to target tumor-initiating traits such as spheroid and colony formation, by altering key signaling pathways that inhibit proliferation and cell viability, induce cell cycle arrest, apoptosis, and senescence in stem cells (as well as differentiated cells). As a result, tumors may be more easily targeted by standard therapies.

The spheroid and colony formation assays measure the potential of cells to self-renew in an anchorage-independent and anchorage-dependent environment, respectively. A number of chemopreventive agents including polysaccharopeptide (PSP, extracted from Turkey tail mushroom *Coriolus versicolor* or Yun-zhi), bone morphogenetic protein (BMP7, a member of the TGF-β superfamily), gamma-tocotrienols (γ-T3, component of vitamin E), genistein (from soy), and epigallocatechin gallate (EGCG, from green tea) have exhibited the ability to inhibit spheroid and colony formation in prostate stem cells (Kobayashi et al. 2011; Luk et al. 2011; Tang et al. 2012; Zhang et al. 2012). EGCG also suppressed the spheroid- and colony-forming ability of human breast CSCs. This was mediated by activation of adenosine monophosphate-activated protein kinase (AMPKα) (Chen et al. 2012). AMPKα serves as an energy sensor in eukaryotic cells and its activation suppresses cell proliferation in malignant cells (Motoshima et al. 2006). Similarly, blueberry polyphenolic acid and genistein inhibited the formation of anchorage-independent spheroid in mammary stem cell cultures. The effects of genistein were mediated by increased expression of phosphatase and tensin homolog (PTEN) in these cells (Montales et al. 2012). Loss of PTEN and overactivation of the PI3K/Akt-mediated proliferative pathway strongly correlate with progression of prostate cancer (Sarkar et al. 2010). Other studies using breast TICs indicated that curcumin and piperine, from turmeric and black pepper, respectively, inhibited spheroid formation by down-regulating Wnt signaling (Kakarala et al. 2010). The Wnt/β-catenin pathways are associated with self-renewal (Li et al. 2011). Studies on brain and colon CSC demonstrate that in the presence of curcumin, signal transducer and activator of transcription 3 (STAT3), a transcription factor, was a major regulator of spheroid and colony formation in vitro and that combination treatment with 5-florouracil and oxaliplatin (FOLFOX) enhanced these effects (Lim et al. 2011; Lin et al. 2011; Yu et al. 2009). Sulforaphane, an organosulfur compound from cruciferous vegetables, blocked colony and spheroid formation in human breast and pancreatic TICs. Nuclear factor kappa-light-chain enhancer of activated B cells (NF-κB), a protein complex that regulates DNA transcription, was downregulated in these studies (Kallifatidis et al. 2011; Li et al. 2010b; Rausch et al. 2010). Recent studies have also begun to focus on the role of fatty acid synthesis, which is downregulated by resveratrol and altered stem properties such as spheroid formation (Pandey et al. 2011).

Using stem cells from a pancreatic tumor, sulforaphane was found to block expression of Nanog, a transcription factor important in self-renewal and differentiation (Srivastava et al. 2011). Similarly, resveratrol prevented expression of genes essential for pluripotency: Sox2, c-Myc, and Oct4 (Shankar et al. 2011). Treatment with curcumin downregulated stem cell markers including Notch1and Hes-1, and pro-survival gene Bcl-XL, in addition to inhibiting NF-κB in pancreatic CSC (Wang et al. 2006). EGCG along with quercetin (a flavonoid found in fruits, vegetables, leaves, and grains) synergized to attenuate TCF/LEF and Gli activities (Tang et al. 2012). TCF/LEF and Gli are involved in Wnt and Sonic Hedgehog signaling; these pathways are essential to supporting a stem cell's self-renewal capacity. Thus, chemopreventive agents present a powerful tool in managing self-renewal of CSCs by modulating various molecular pathways.

#### 8.3.2.2 Differentiation of Cancer Stem Cells

Selective targeting of CSCs in order to avoid normal healthy cells has been challenging; therefore, efforts in chemoprevention have focused on methods of driving CSC into differentiated lineages. The concept of "differentiation therapy" was first proposed in 1961 with the hypothesis that CSCs, upon differentiation, would give rise to progenitors and subsequently differentiated cancer cells that would gradually deplete, resulting in tumor regression (Rane et al. 2012). The proof for this premise came from the clinical trial for promyelocytic leukemia using all-trans retinoic acid (ATRA), where the majority of patients underwent remission and blast cell differentiation was demonstrated (Huang et al. 1988). Based on the success achieved in treating leukemia, there has been a substantial effort in screening chemoprevention compounds that induce differentiation of normal and solid organ tumor-derived stem cells.

Chemoprevention-mediated differentiation of various normal stem cell populations is well documented. Sulforaphane induced human promyelocytic cells to differentiate into granulocytes and macrophages via activity of PI3K and PKC (Fimognari et al. 2008). Our lab has shown that vitamin D stimulates murine prostate progenitor/stem cells (PrP/SCs) in an IL-1α-dependent manner toward a luminal cell fate (Maund et al. 2011). Other labs have also found vitamin D to stimulate differentiation of human mesenchymal stem cells (hMSC) into osteoblasts through signaling pathways involving regulation of c-Myc by BMP2, or HGF by upregulation of the vitamin D receptor (D'Ippolito et al. 2002; Kuske et al. 2011; Piek et al. 2010).

Vitamin D, with the help of retinoic acid, also increased the differentiation of adipose-derived stem cells (Malladi et al. 2006). Resveratrol had similar effects on hMSC (Dai et al. 2007), as well as inducing pluripotent stem cells to differentiate into functional osteoclasts (Kao et al. 2010). Similarly, fatty alcohol derivatives of resveratrol were able to stimulate the maturation of neuronal stem cells (Hauss et al. 2007), indicating that many methods of differentiation are possible with chemoprevention. These studies highlight the role of chemoprevention in driving cells toward a more differentiated fate.

Chemopreventive intervention has also been explored as a method to drive CSCs down the path of differentiation. Stem cell antigen (Sca-1) is a cell-surface marker that has been used to enrich the stem/progenitor-like subpopulation. Expression of the Sca-1 surface marker was significantly reduced on prostate CSCs by treatment with a prolactin inhibitor (Rouet et al. 2010). Similarly, an α(v)-integrin antagonist, GLPG0187, reduced the percentage of ALDH-positive stem cells isolated from a population of PC-3 human prostate cancer cell line (van der Horst et al. 2011). Furthermore, treatment of prostate stem cells with γ-T3, PSP, and genistein prevented expression of the stemness markers, CD133 and CD44 (Ling et al. 2011; Luk et al. 2011). These results indicate that cells may have been stimulated to leave their stemlike state in favor of a terminal mature fate.

Differentiation of CSCs caused by treatment with chemopreventive agents has the potential to bring the heterogeneous tumor population to a more consistent differentiated phenotype. The ability to treat a more homogenously differentiated tumor has led to greater success in eliminating cancer cells with standard treatment. However, recurrence is still common for tumors in which cells remain heterogeneous. In an effort to halt progression of prostatic tumor growth, stem cells continue to be a target of chemoprevention research.

#### 8.3.2.3 Growth Arrest of Cancer Stem Cells

Studies from other systems have shed light on pathways that may affect apoptosis as well as proliferation and senescence. Viability of human prostate stem cells was compromised by exposure to γ-T3 (Luk et al. 2011), and senescence was induced by BMP7 to activate the p38 MAP kinase pathway (Kobayashi et al. 2011). EGCG reduced cell viability and inhibited proliferation of prostate and pancreatic CSCs (Tang et al. 2010, 2012). Vitamin D has been shown not only to inhibit proliferation but also to target stem cells of both prostate and bone by arresting cells in the $G_0/G_1$ phase or by inducing cellular senescence (Artaza et al. 2010; Li et al. 2009). Synergism between vitamin D and genistein has also been reported to cause these effects in prostate cancer cell lines (Rao et al. 2002, 2004); however, this synergism has not been tested in CSCs. Additionally, vitamin D-mediated growth arrest, differentiation, and $G_0/G_1$ arrest are shown to be mediated by IL-1α in mouse prostate progenitor cells (Maund and Cramer 2010).

Curcumin treatment of multiple brain tumor cell cultures resulted in decreased viability, cell cycle arrest in $G_2/M$ phase along with reduction of $CD133^+$ stem cells;

attenuation of STAT3 and suppression of Notch signaling were shown to mediate these effects (Lim et al. 2011). Cucurbitacins, tetracycline triterpenoids initially identified in cucumbers, are inhibitors of the Jak/STAT pathway. It was reported to reduce $CD133^{+}$ medulloblastoma CSC by reducing phosphorylation of STAT3 (Chang et al. 2012). Similarly, plant derivatives parthenolide and andrographolide were selectively toxic to multiple myeloma CSC compared to non-tumorigenic cells (Gunn et al. 2011). Resveratrol and sulforaphane also induced apoptosis in leukemia stem cells (Fimognari et al. 2008; Hu et al. 2012) and had similar effects in breast, bone, and pancreas by altering expression of apoptosis markers like BNIP3, DAPK2, RANKL, XIAP, caspase 3/7, and Bcl-2 (He et al. 2010; Nishikawa et al. 2009; Pandey et al. 2011; Shankar et al. 2011; Srivastava et al. 2011). In both human and mouse colon stem cells, nonsteroidal anti-inflammatory drugs (NSAIDS) like sulindac and the curcumin analog GO-Y030 have inhibited proliferation by inducing apoptosis and reducing cell viability, either by inhibiting the β-catenin and Wnt signaling pathway (Qiu et al. 2010) or by reducing expression of STAT3 (Lin et al. 2011) in colon CSCs. These pathways could potentially be active targets in prostate cancer and warrant further investigation.

#### 8.3.2.4 Altering Genetic, Epigenetic, and Signaling Pathways

Chemopreventive agents such as sulforaphane, curcumin, genistein, EGCG, quercetin, and lycopene have been shown to regulate cellular epigenetics and various signaling pathways (Vanden Berghe 2012). Epigenetic regulation by DNA methylation, histone modifications, and microRNA can contribute to the process of carcinogenesis by modifying proteins and signaling networks. The reversible nature of epigenetic modifications makes them good targets for chemoprevention. There is substantial literature available on the epigenetic impact of dietary polyphenols in cancer chemoprevention (Izzotti et al. 2012; Li et al. 2010a; Vanden Berghe 2012; Vira et al. 2012). However, there are few studies evaluating these mechanisms in CSC. Yu et al. demonstrated the elimination of colon CSC by the combination of curcumin and FOLFOX. These results were attributed to hypomethylation of the EGFR promoter and alterations in the levels of DNA methyltransferase 1 (Yu et al. 2009) indicating a possible role for epigenetic changes stimulated by chemoprevention. Similar to oncogenes and tumor suppressor genes, microRNAs (miRNAs) have been shown to have cancer promoting and suppressive functions. As many as 145 miRNAs involved in carcinogenesis mechanisms have been reported to be modulated by natural and synthetic agents, either individually or in combination (Izzotti et al. 2012). Kanwar et al. reported that fluorinated curcumin (CDF) decreases the expression of EZH2 (histone H3K27m3 thrimethyl transferase) and overexpression of miRNAs let-7a, b, c; miR-26a; mir-101; miR-146a; and mir-200b, c in colon CSC population (Kanwar et al. 2011). Similarly, in pancreatic CSC, CDF decreased EZH2, CD44, and Nanog and upregulated miR-7, miR-26a, and miR-200b (Bao et al. 2012). These studies highlight the role of chemoprevention-mediated epigenetic modulation in restricting the CSC population. Various natural and synthetic chemopreventive agents have demonstrated regulation of miRNAs in prostate cancer cells, but their role in prostate CSCs remains to be elucidated (reviewed in Maugeri-Sacca et al. 2012).

#### 8.3.2.5 Sensitizing Cancer Stem Cells to Traditional Therapies

A major hurdle in targeting stem cells is their innate resistance to toxic chemotherapies. Upregulation of multidrug resistance transporters (MDRT), specifically multidrug resistance protein 1 (MDR1), responsible for effluxing toxins, is common in stem cell populations making them a very challenging target. Stem cells that are not cleared by treatment go on to repopulate a tumor following remission (Fig. 8.3). Remarkably, curcumin treatment reduced MDRT/MDR1 expression in stem cells in the brain (Fong et al. 2010), ovaries, breast (Limtrakul et al. 2007), and blood (Anuchapreeda et al. 2006), making them more susceptible to therapy. Natural curcumin and a curcumin analog, difluorinated-curcumin (CDF), were also able to downregulate CD44 and CD166 in human colon stem cells. This effect was enhanced when combined with FOLFOX (Kanwar et al. 2011; Yu et al. 2009), both of which are strong DNA-damaging agents that may also hinder kinase activity involved in genomic repair. Sulforaphane was also identified to promote sensitivity to gemcitabine, doxorubicin, and 5-FU in stem cells enriched from prostate and pancreatic cancer cell lines (Kallifatidis et al. 2011). Sulforaphane reduced ALDH activity in human pancreatic CSC when co-treated with quercetin (Zhou et al. 2010). Similarly, sulforaphane augmented the elimination of pancreatic CSC when combined with sorafenib; this was demonstrated to be due to increased DNA fragmentation and apoptosis and decreased cell proliferation and angiogenesis and by downregulation of NF-κB activity. Additionally, these treatments were selectively toxic to the CSC but not the nonmalignant cells (Rausch et al. 2010). These reports suggest that chemopreventive agents can be an excellent choice for adjuvant therapy.

### *8.3.3 Preventing Tumor Growth In Vivo*

CSC form a rare cell population in tumors, though they have the ability to give rise to tumors, even when as few as one cell is xenografted in vivo (Quintana et al. 2008). Most in vivo chemoprevention studies have not utilized such a limiting dilution model; nonetheless, they provide an insight into the long-term antitumorigenic effects of these agents. Gamma-tocotrienol and PSP inhibited in vivo tumor initiation of human prostate cancer cell lines enriched for stem cells (Luk et al. 2011). Genistein prevented in vivo prostate tumor growth in mice when stem cells were pretreated prior to transplantation (Zhang et al. 2012). Colon and breast stem cells from both human and mouse similarly responded to nonsteroidal anti-inflammatory drugs (NSAIDs) and sulforaphane, respectively (Li et al. 2010b; Qiu et al. 2010). Sulforaphane also abrogated tumor growth of $ALDH^+$ breast CSCs in vivo by downregulating the Wnt/β-catenin pathway (Li et al. 2010b). $APC^{Min/+}$ mice develop

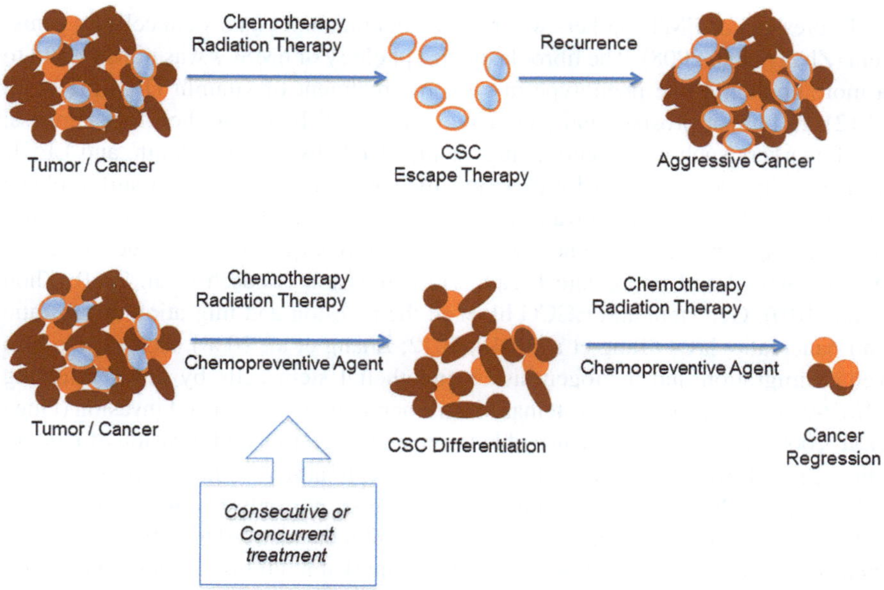


**Fig. 8.3** Conventional chemo- and radiation therapy target rapidly growing mature cells, leaving behind the quiescent CSC. These residual CSCs are thought to be the cause of cancer recurrence. Chemopreventive agents can cause differentiation of CSC, thereby making tumors more targetable

spontaneous intestinal adenomas and are a good in vivo model for studying colon cancer. Sulindac effectively prevented formation of polyps and colon cancer by inducing apoptosis in Lrg5$^+$ colon stem cells (Qiu et al. 2010). All of these functions, stimulated by a naturally occurring chemopreventive agent, impede tumor initiation by targeting CSC.

### *8.3.4 Preventing Metastasis*

Metastasis leads to a dramatically reduced survival rate of patients. Chemoprevention provides the opportunity to target prostate cancer at any stage of development, including the prevention of metastasis. The epithelial to mesenchymal transition (EMT) is a process by which tumor cells change their phenotype from a stationary epithelial cell to a more motile mesenchymal cell and is believed to be the initial transition to a metastatic phenotype. Protein markers and motility studies, for invasion and migration, help in identifying this transition. Chemopreventive agents have been shown to prevent such transformations.

Expression of EMT markers was reduced in human prostate stem cells by genistein (Zhang et al. 2008). The fibroblastic morphology of hMSCs was transformed to a more epithelial-like phenotype in vitro by treatment of vitamin D (Klotz et al. 2012). In human prostate and pancreatic cancer cell lines enriched for stemness, EMT markers such as vimentin, slug, snail, ZEB1, twist-1, β-catenin, and LEF1/TCF activity were reduced by exposure to EGCG, resveratrol, and sulforaphane (Shankar et al. 2011; Srivastava et al. 2011; Tang et al. 2010). Additionally, sulforaphane was also reported to synergize with sorafenib or quercetin to reduce the EMT marker twist-2 and upregulate E-cadherin expression (Rausch et al. 2010; Zhou et al. 2010). Genistein and EGCG blocked the invasion and migration of prostatic and pancreatic TICs (Tang et al. 2010, 2012; Zhang et al. 2008). EGCG also prevented migration and angiogenesis of endothelial stem cells by downregulating MMP-9, a secreted metalloproteinase instrumental in migration and invasion (Ohga et al. 2009). Resveratrol was also able to prevent migration and invasion of pancreatic stem cells (Shankar et al. 2011). Furthermore, treatment of human prostate stem cells with GLPG0187, an α(v)-integrin antagonist, not only resulted in increased expression of E-cadherin and reduction of vimentin but also led to reduction in bone metastasis (van der Horst et al. 2011). The remarkable ability of these chemopreventive agents to deter the process of metastasis proves to be one of the many promising directions of the field. Further understanding of these chemopreventive agents and their role in preventing metastasis may lead to the development of new therapeutics for prostate cancer.

## 8.4 Issues and Limitations of Chemoprevention

Currently there are over 2,000 interventional clinical trials that are focused on either evaluating the prevalence of CSCs in various cancers or measuring them as an outcome of chemotherapy or radiation therapy. On the other hand, there are currently about 200 clinical trials evaluating dietary compounds for cancer chemoprevention, but none seems to be assessing the effects on CSCs (http://www.clinicaltrials.gov/). Despite data from various research laboratories demonstrating that chemopreventive agents make stem cells more susceptible to standard therapies, thereby suggesting their promising role for prophylactic or adjuvant therapy, there is significant skepticism in clinics.

CSCs are being increasingly recognized as the cause of recurrence and metastasis, but there is a lack of knowledge about "what defines CSC." Additionally, the concept of tissue-specific stem cells and the role of tissue microenvironment in modulating functional and phenotypic nature of CSCs prevent generalization of concepts formulated in one cancer to other cancer types. In addition to the chemopreventive agents discussed in this chapter, several other compounds like lycopene, silibinin, 3,3′-diindolylmethane (DIM), pomegranate, fisetin, lupeol (Khan et al. 2010; Sarkar et al. 2010; Venkateswaran and Klotz 2010), and mangostin (Johnson et al. 2011) have been characterized in a variety of cancers (including prostate cancer), but remain unevaluated for their CSC-specific response. This incomplete understanding of tissue-specific cancers, CSC regulation, and possible tissue-specific role of natural chemopreventive agents could potentially lead to unsuccessful outcome of chemoprevention trials.

### 8.4.1 *Limitations of In Vitro and In Vivo Model*

The in vitro colony and serial spheroid formation and 3D cultures are excellent surrogate assays for measuring the self-renewal potential of NSCs and CSCs/TICs. Additionally, evaluating these properties in the context of treatment with various natural and synthetic agents helps identify promising compounds for therapeutics and chemoprevention, as well as understanding their mode of action (Table 8.1). However, these are isolated systems and results might not always extend to a whole body system. Thus, it is imperative to conduct in vivo studies. As discussed earlier, a reduced or delayed tumor growth was reported when human prostate CSCs were pretreated with vitamin E, PSP, or genistein prior to implanting in mice. These studies hint at the prophylactic potential of chemopreventive agents. On the other hand, Rausch et al. reported the effectiveness of addition of sulforaphane to sorafenib in eliminating pancreatic CSCs and restricting tumor growth in vivo, which shows the potential of a natural compound to be used in adjuvant therapy (Rausch et al. 2010). But one must cautiously interpret these results for two reasons: (1) In in vivo models animals are inoculated with a substantial number of CSCs, and (2) these in vivo experiments are almost always conducted in immunocompromised animals. As previously discussed, CSCs or TICs likely form a very small proportion of tumors. Accordingly, it would be more informative to conduct "limiting dilution" experiments in conjunction with pretreatment or combinational therapeutic intervention to eliminate CSCs by using chemopreventive agents. Alternatively, a mixture of CSC and non-CSCs (in combination with tumor-associated mesenchymal cells, Chap. 7) should be inoculated to more accurately mimic the native tumor environment. The immune system is an important aspect of clinical cancer. The in vivo experiments in immunocompromised murine cancer models provide an insight into the effectiveness of chemopreventive agents in a whole body system and allow for the use of human/clinical samples, but they ignore the immunological manifestations that could potentially alter the response to these chemopreventive compounds.

A number of chemoprevention studies have been conducted in murine models that spontaneously develop prostate cancer (reviewed in Lamb and Zhang 2005). However, most of these systems do not accurately mimic the sequence of events in prostate cancer progression. Additionally, conventional in vivo studies involve heterotopic implantation (i.e., subcutaneous, intravenous, intracardiac) of CSCs in immunocompromised murine models. These models address the question of whether CSCs can proliferate and/or metastasize in vivo, but the lack of their histological similarity with human pathology makes it difficult to interpret these results in terms of clinical cancer. Chemoprevention is a very promising method to inhibit cancer progression, but the skepticism in the field could be a manifestation of the current scarcity of accurate models to conduct these studies.

There is a significant impetus in identifying CSC (see Chaps. 1, 2 and 3) and developing rodent models that mimic human cancer (see Chap. 9). Isolating the TICs/CSCs from these animals and evaluating chemoprevention in syngeneic models will result in a better comprehension of the effectiveness of these compounds in blocking carcinogenesis. Our lab has developed an elegant model that addresses these issues where we have shown complete multi-lineage differentiation of a single cell clonal population, derived from murine prostate, in tissue recombinant experiments in vivo (Barclay et al. 2005). Using various gene expression and deletion constructs, the role of specific genes in formation of neoplastic lesions compared to normal prostate architecture can be evaluated. We demonstrated that loss of TGFβ-1 activating kinase (Tak1) in murine prostate progenitor cells led to formation of PIN and even carcinoma (Wu et al. 2012). Integration of chemoprevention in these models will help us evaluate the usefulness of specific agents in different phases of carcinogenesis. It would also be most interesting to develop mice in which certain genes, under the control of tissue-specific stem cell promoter, could be exclusively altered in the stem cells of adult animals. This would address the limitations of many transgenic models, i.e. (1) the lack or overexpression of tumor suppressor or oncogenes from embryonic stage and (2) no concerns of embryonic lethality since it would be an intact animal, unless treated to induce gene alteration. Although no stem-cell-specific promoters in the prostate are known yet, development of such a model would not only help in understanding cancer progression but would be valuable in developing preventive and therapeutic compounds.

### 8.4.2 Timing and Dosage of Chemopreventive Intervention

Cancer chemoprevention during the early phases of carcinogenesis is a viable approach for controlling most cancers; however, it might become a daunting and unmanageable reality as cancer advances. This is most likely due to the increasing number of genetic mutations with time. It was reported that vitamin D was maximally effective in preventing formation of prostatic intraepithelial neoplasia (PIN) in the *NKx3.1; Pten* mutant mice when administered prior to, rather than subsequent to, the initial occurrence of PIN (Banach-Petrosky et al. 2006). Thus, it is appropriate to suggest that specific recommendation for cancer prevention must be based on how far along a patient is on the scale of cancer progression. The models discussed above and in Chap. 9 can help in teasing out these details and developing efficient methods for cancer prevention.

The use of ATRA was one of the pioneering and promising trials supporting chemoprevention or differentiation therapy. Subsequently, ATRA and retinoic acid analogs have been tested in prostate cancer trials, but with limited to no success (Trump et al. 1997). One of the criticisms for such failed trials is the use of considerable variations in dosage. Most chemopreventive compounds, like retinoic acid (Fong et al. 1993), exhibit a dose-dependent biphasic effect, and their inappropriate use could result in unexpected toxic outcomes. Hence, it seems probable that chemopreventive agents that are effective in controlling carcinogenesis at lower dosages might be ineffective or even pro-cancerous at elevated levels.

## 8.5 Discussion

While impressive progress has been made in the fields of cancer diagnosis and treatment, cancer remains a serious public health concern. In addition to the toxic side effects of cancer therapy, one of the many challenges that obstruct eradication of cancer is the development of resistance to traditional therapies by cancer cells that ultimately lead to recurrence and death. This resistance and relapse is now being attributed to the rare cancer-initiating stem cell population in tumors. Studies in the past few decades have revealed that certain natural and synthetic dietary compounds can help in cancer management by regulating self-renewal and differentiation of CSC. Thus, chemoprevention emerges as an important tool in controlling cancer progression.

Cancer chemoprevention is believed to be a viable approach to slow or prevent the process of carcinogenesis for most solid organ malignancies. Drawing inspiration from ancient cultures, rapid progress is being made in identifying new chemopreventive agents. The efficacy of many of these agents has been tested in different cancer types. Recently, their role in modulating the chemotherapy-resistant CSC has gained impetus. In addition to low toxicity, the advantages of using chemopreventive agents is that they have multifaceted impacts on tumor cells and CSCs. It is important that these compounds, individually or in combination, specifically target tumor-initiating or CSC and not compromise healthy stem cells in the body. Nonetheless, there is a growing concern about unexpected toxicities as a result of extended chemopreventive regimes. The use of nanotechnology has been suggested and is currently being explored in order to increase bioavailability, improve sustained delivery and reduce toxicities. Advent of more clinically relevant in vitro and in vivo models is expected to improve screening of chemopreventive agents that can subsequently be incorporated in prophylactic or adjuvant therapy. While the molecular mechanisms associated with tissue-specific cancers and CSC are still under investigation, we hope that lessons learned from one system can be extended cautiously to another. Moreover, the outcome of chemoprevention could not only be associated with tumor stage and ongoing therapy but may also be altered by race, genetics, geographical location, and diet. Thus, chemoprevention can become a success story by identifying the right patient population, the appropriate time and dose for interventions, and administration of the suitable agent.

## References

Abate-Shen C, Shen MM (2000) Molecular genetics of prostate cancer. Genes Dev 14:2410–2434

Anuchapreeda S, Thanarattanakorn P, Sittipreechacharn S, Tima S, Chanarat P, Limtrakul P (2006) Inhibitory effect of curcumin on MDR1 gene expression in patient leukemic cells. Arch Pharm Res 29:866–873

Artaza JN, Sirad F, Ferrini MG, Norris KC (2010) 1,25(OH)2vitamin D3 inhibits cell proliferation by promoting cell cycle arrest without inducing apoptosis and modifies cell morphology of mesenchymal multipotent cells. J Steroid Biochem Mol Biol 119:73–83

Banach-Petrosky W, Ouyang X, Gao H, Nader K, Ji Y, Suh N, DiPaola RS, Abate-Shen C (2006) Vitamin D inhibits the formation of prostatic intraepithelial neoplasia in Nkx3.1;Pten mutant mice. Clin Cancer Res 12:5895–5901

Bao B, Ali S, Banerjee S, Wang Z, Logna F, Azmi AS, Kong D, Ahmad A, Li Y, Padhye S et al (2012) Curcumin analogue CDF inhibits pancreatic tumor growth by switching on suppressor microRNAs and attenuating EZH2 expression. Cancer Res 72:335–345

Barclay WW, Woodruff RD, Hall MC, Cramer SD (2005) A system for studying epithelial-stromal interactions reveals distinct inductive abilities of stromal cells from benign prostatic hyperplasia and prostate cancer. Endocrinology 146:13–18

Chang CJ, Chiang CH, Song WS, Tsai SK, Woung LC, Chang CH, Jeng SY, Tsai CY, Hsu CC, Lee HF et al (2012) Inhibition of phosphorylated STAT3 by cucurbitacin I enhances chemoradiosensitivity in medulloblastoma-derived cancer stem cells. Childs Nerv Syst 28:363–373

Chen D, Pamu S, Cui Q, Chan TH, Dou QP (2012) Novel epigallocatechin gallate (EGCG) analogs activate AMP-activated protein kinase pathway and target cancer stem cells. Bioorg Med Chem 20:3031–3037

D'Ippolito G, Schiller PC, Perez-stable C, Balkan W, Roos BA, Howard GA (2002) Cooperative actions of hepatocyte growth factor and 1,25-dihydroxyvitamin D3 in osteoblastic differentiation of human vertebral bone marrow stromal cells. Bone 31:269–275

Dai Z, Li Y, Quarles LD, Song T, Pan W, Zhou H, Xiao Z (2007) Resveratrol enhances proliferation and osteoblastic differentiation in human mesenchymal stem cells via ER-dependent ERK1/2 activation. Phytomedicine 14:806–814

Denissova NG, Nasello CM, Yeung PL, Tischfield JA, Brenneman MA (2012) Resveratrol protects mouse embryonic stem cells from ionizing radiation by accelerating recovery from DNA strand breakage. Carcinogenesis 33:149–155

Dick JE (2008) Stem cell concepts renew cancer research. Blood 112:4793–4807

Donnenberg VS, Donnenberg AD (2005) Multiple drug resistance in cancer revisited: the cancer stem cell hypothesis. J Clin Pharmacol 45:872–877

Epifano F, Genovese S, Miller R, Majumdar AP (2012) Auraptene and its effects on the re-emergence of colon cancer stem cells. Phytother Res Jul 4. doi: 10.1002/ptr.4773.

Fimognari C, Lenzi M, Cantelli-Forti G, Hrelia P (2008) Induction of differentiation in human promyelocytic cells by the isothiocyanate sulforaphane. In Vivo 22:317–320

Fong CJ, Sutkowski DM, Braun EJ, Bauer KD, Sherwood ER, Lee C, Kozlowski JM (1993) Effect of retinoic acid on the proliferation and secretory activity of androgen-responsive prostatic carcinoma cells. J Urol 149:1190–1194

Fong D, Yeh A, Naftalovich R, Choi TH, Chan MM (2010) Curcumin inhibits the side population (SP) phenotype of the rat C6 glioma cell line: towards targeting of cancer stem cells with phytochemicals. Cancer Lett 293:65–72

Gunn EJ, Williams JT, Huynh DT, Iannotti MJ, Han C, Barrios FJ, Kendall S, Glackin CA, Colby DA, Kirshner J (2011) The natural products parthenolide and andrographolide exhibit anti-cancer stem cell activity in multiple myeloma. Leuk Lymphoma 52:1085–1097

Hauss F, Liu J, Michelucci A, Coowar D, Morga E, Heuschling P, Luu B (2007) Dual bioactivity of resveratrol fatty alcohols: differentiation of neural stem cells and modulation of neuroinflammation. Bioorg Med Chem Lett 17:4218–4222

He X, Andersson G, Lindgren U, Li Y (2010) Resveratrol prevents RANKL-induced osteoclast differentiation of murine osteoclast progenitor RAW 264.7 cells through inhibition of ROS production. Biochem Biophys Res Commun 401:356–362

Hu L, Cao D, Li Y, He Y, Guo K (2012) Resveratrol sensitized leukemia stem cell-like KG-1a cells to cytokine-induced killer cells-mediated cytolysis through NKG2D ligands and TRAIL receptors. Cancer Biol Ther 13:516–526

Huang ME, Ye YC, Chen SR, Chai JR, Lu JX, Zhoa L, Gu LJ, Wang ZY (1988) Use of all-trans retinoic acid in the treatment of acute promyelocytic leukemia. Blood 72:567–572

Ito T, Zimdahl B, Reya T (2012) aSIRTing control over cancer stem cells. Cancer Cell 21:140–142

Izrailit J, Reedijk M (2012) Developmental pathways in breast cancer and breast tumor-initiating cells: therapeutic implications. Cancer Lett 317:115–126

Izzotti A, Cartiglia C, Steele VE, De Flora S (2012) MicroRNAs as targets for dietary and pharmacological inhibitors of mutagenesis and carcinogenesis. Mutat Res 751(2):287–303

Johnson JJ, Petiwala SM, Syed DN, Rasmussen JT, Adhami VM, Siddiqui IA, Kohl AM, Mukhtar H (2011) Alpha-Mangostin, a xanthone from mangosteen fruit, promotes cell cycle arrest in prostate cancer and decreases xenograft tumor growth. Carcinogenesis 33:413–419

Kakarala M, Brenner DE, Korkaya H, Cheng C, Tazi K, Ginestier C, Liu S, Dontu G, Wicha MS (2010) Targeting breast stem cells with the cancer preventive compounds curcumin and piperine. Breast Cancer Res Treat 122:777–785

Kallifatidis G, Labsch S, Rausch V, Mattern J, Gladkich J, Moldenhauer G, Buchler MW, Salnikov AV, Herr I (2011) Sulforaphane increases drug-mediated cytotoxicity toward cancer stem-like cells of pancreas and prostate. Mol Ther 19:188–195

Kanwar SS, Yu Y, Nautiyal J, Patel BB, Padhye S, Sarkar FH, Majumdar AP (2011) Difluorinated-curcumin (CDF): a novel curcumin analog is a potent inhibitor of colon cancer stem-like cells. Pharm Res 28:827–838

Kao CL, Tai LK, Chiou SH, Chen YJ, Lee KH, Chou SJ, Chang YL, Chang CM, Chen SJ, Ku HH et al (2010) Resveratrol promotes osteogenic differentiation and protects against dexamethasone damage in murine induced pluripotent stem cells. Stem Cells Dev 19:247–258

Kavalerchik E, Goff D, Jamieson CH (2008) Chronic myeloid leukemia stem cells. J Clin Oncol 26:2911–2915

Khan N, Adhami VM, Mukhtar H (2010) Apoptosis by dietary agents for prevention and treatment of prostate cancer. Endocr Relat Cancer 17:R39–R52

Klein EA (2005) Can prostate cancer be prevented? Nat Clin Pract Urol 2:24–31

Klotz B, Mentrup B, Regensburger M, Zeck S, Schneidereit J, Schupp N, Linden C, Merz C, Ebert R, Jakob F (2012) 1,25-dihydroxyvitamin D3 treatment delays cellular aging in human mesenchymal stem cells while maintaining their multipotent capacity. PLoS One 7:e29959

Kobayashi A, Okuda H, Xing F, Pandey PR, Watabe M, Hirota S, Pai SK, Liu W, Fukuda K, Chambers C et al (2011) Bone morphogenetic protein 7 in dormancy and metastasis of prostate cancer stem-like cells in bone. J Exp Med 208:2641–2655

Kuske B, Savkovic V, zur Nieden NI (2011) Improved media compositions for the differentiation of embryonic stem cells into osteoblasts and chondrocytes. Methods Mol Biol 690:195–215

Lamb DJ, Zhang L (2005) Challenges in prostate cancer research: animal models for nutritional studies of chemoprevention and disease progression. J Nutr 135:3009S–3015S

Li H, Tang DG (2011) Prostate cancer stem cells and their potential roles in metastasis. J Surg Oncol 103:558–562

Li J, Fleet JC, Teegarden D (2009) Activation of rapid signaling pathways does not contribute to 1 alpha,25-dihydroxyvitamin D3-induced growth inhibition of mouse prostate epithelial progenitor cells. J Cell Biochem 107:1031–1036

Li Y, Kong D, Wang Z, Sarkar FH (2010a) Regulation of microRNAs by natural agents: an emerging field in chemoprevention and chemotherapy research. Pharm Res 27:1027–1041

Li Y, Zhang T, Korkaya H, Liu S, Lee HF, Newman B, Yu Y, Clouthier SG, Schwartz SJ, Wicha MS et al (2010b) Sulforaphane, a dietary component of broccoli/broccoli sprouts, inhibits breast cancer stem cells. Clin Cancer Res 16:2580–2590

Li Y, Wicha MS, Schwartz SJ, Sun D (2011) Implications of cancer stem cell theory for cancer chemoprevention by natural dietary compounds. J Nutr Biochem 22:799–806

Lim KJ, Bisht S, Bar EE, Maitra A, Eberhart CG (2011) A polymeric nanoparticle formulation of curcumin inhibits growth, clonogenicity and stem-like fraction in malignant brain tumors. Cancer Biol Ther 11:464–473

Limtrakul P, Chearwae W, Shukla S, Phisalphong C, Ambudkar SV (2007) Modulation of function of three ABC drug transporters, P-glycoprotein (ABCB1), mitoxantrone resistance protein (ABCG2) and multidrug resistance protein 1 (ABCC1) by tetrahydrocurcumin, a major metabolite of curcumin. Mol Cell Biochem 296:85–95

Lin L, Liu Y, Li H, Li PK, Fuchs J, Shibata H, Iwabuchi Y, Lin J (2011) Targeting colon cancer stem cells using a new curcumin analogue, GO-Y030. Br J Cancer 105:212–220

Ling MT, Luk SU, Al-Ejeh F, Khanna KK (2011) Tocotrienol as a potential anticancer agent. Carcinogenesis 33:233–239

Luk SU, Lee TK, Liu J, Lee DT, Chiu YT, Ma S, Ng IO, Wong YC, Chan FL, Ling MT (2011) Chemopreventive effect of PSP through targeting of prostate cancer stem cell-like population. PLoS One 6:e19804

Maitland NJ, Collins AT (2008) Prostate cancer stem cells: a new target for therapy. J Clin Oncol 26:2862–2870

Malladi P, Xu Y, Yang GP, Longaker MT (2006) Functions of vitamin D, retinoic acid, and dexamethasone in mouse adipose-derived mesenchymal cells. Tissue Eng 12:2031–2040

Maugeri-Sacca M, Coppola V, Bonci D, De Maria R (2012) MicroRNAs and prostate cancer: from preclinical research to translational oncology. Cancer J 18:253–261

Maund SL, Cramer SD (2010) The tissue-specific stem cell as a target for chemoprevention. Stem Cell Rev 7:307–314

Maund SL, Barclay WW, Hover LD, Axanova LS, Sui G, Hipp JD, Fleet JC, Thorburn A, Cramer SD (2011) Interleukin-1alpha mediates the antiproliferative effects of 1,25-dihydroxyvitamin D3 in prostate progenitor/stem cells. Cancer Res 71:5276–5286

Montales MT, Rahal OM, Kang J, Rogers TJ, Prior RL, Wu X, Simmen RC (2012) Repression of mammosphere formation of human breast cancer cells by soy isoflavone genistein and blueberry polyphenolic acids suggests diet-mediated targeting of cancer stem-like/progenitor cells. Carcinogenesis 33:652–660

Motoshima H, Goldstein BJ, Igata M, Araki E (2006) AMPK and cell proliferation—AMPK as a therapeutic target for atherosclerosis and cancer. J Physiol 574:63–71

Nishikawa T, Tsuno NH, Tsuchiya T, Yoneyama S, Yamada J, Shuno Y, Okaji Y, Tanaka J, Kitayama J, Takahashi K et al (2009) Sulforaphane stimulates activation of proapoptotic protein bax leading to apoptosis of endothelial progenitor cells. Ann Surg Oncol 16:534–543

Ohga N, Hida K, Hida Y, Muraki C, Tsuchiya K, Matsuda K, Ohiro Y, Totsuka Y, Shindoh M (2009) Inhibitory effects of epigallocatechin-3 gallate, a polyphenol in green tea, on tumor-associated endothelial cells and endothelial progenitor cells. Cancer Sci 100:1963–1970

Pandey PR, Okuda H, Watabe M, Pai SK, Liu W, Kobayashi A, Xing F, Fukuda K, Hirota S, Sugai T et al (2011) Resveratrol suppresses growth of cancer stem-like cells by inhibiting fatty acid synthase. Breast Cancer Res Treat 130:387–398

Piek E, Sleumer LS, van Someren EP, Heuver L, de Haan JR, de Grijs I, Gilissen C, Hendriks JM, van Ravestein-van Os RI, Bauerschmidt S et al (2010) Osteo-transcriptomics of human mesenchymal stem cells: accelerated gene expression and osteoblast differentiation induced by vitamin D reveals c-MYC as an enhancer of BMP2-induced osteogenesis. Bone 46:613–627

Qiu W, Wang X, Leibowitz B, Liu H, Barker N, Okada H, Oue N, Yasui W, Clevers H, Schoen RE et al (2010) Chemoprevention by nonsteroidal anti-inflammatory drugs eliminates oncogenic intestinal stem cells via SMAC-dependent apoptosis. Proc Natl Acad Sci USA 107: 20027–20032

Quintana E, Shackleton M, Sabel MS, Fullen DR, Johnson TM, Morrison SJ (2008) Efficient tumour formation by single human melanoma cells. Nature 456:593–598

Rane JK, Pellacani D, Maitland NJ (2012) Advanced prostate cancer—a case for adjuvant differentiation therapy. Nat Rev Urol 9(10):595–602

Rao A, Woodruff RD, Wade WN, Kute TE, Cramer SD (2002) Genistein and vitamin D synergistically inhibit human prostatic epithelial cell growth. J Nutr 132:3191–3194

Rao A, Coan A, Welsh JE, Barclay WW, Koumenis C, Cramer SD (2004) Vitamin D receptor and p21/WAF1 are targets of genistein and 1,25-dihydroxyvitamin D3 in human prostate cancer cells. Cancer Res 64:2143–2147

Rausch V, Liu L, Kallifatidis G, Baumann B, Mattern J, Gladkich J, Wirth T, Schemmer P, Buchler MW, Zoller M et al (2010) Synergistic activity of sorafenib and sulforaphane abolishes pancreatic cancer stem cell characteristics. Cancer Res 70:5004–5013

Ricci-Vitiani L, Fabrizi E, Palio E, De Maria R (2009) Colon cancer stem cells. J Mol Med (Berl) 87:1097–1104

Rouet V, Bogorad RL, Kayser C, Kessal K, Genestie C, Bardier A, Grattan DR, Kelder B, Kopchick JJ, Kelly PA et al (2010) Local prolactin is a target to prevent expansion of basal/stem cells in prostate tumors. Proc Natl Acad Sci USA 107:15199–15204

Sarkar FH, Li Y, Wang Z, Kong D (2010) Novel targets for prostate cancer chemoprevention. Endocr Relat Cancer 17:R195–R212

Shankar S, Nall D, Tang SN, Meeker D, Passarini J, Sharma J, Srivastava RK (2011) Resveratrol inhibits pancreatic cancer stem cell characteristics in human and KrasG12D transgenic mice by inhibiting pluripotency maintaining factors and epithelial-mesenchymal transition. PLoS One 6:e16530

Siegel C (2012) Re: Prostate cancer: prediction of biochemical failure after external-beam radiation therapy—Kattan nomogram and endorectal MR imaging estimation of tumor volume. J Urol 188:432–433

So JY, Lee HJ, Smolarek AK, Paul S, Wang CX, Maehr H, Uskokovic M, Zheng X, Conney AH, Cai L et al (2011) A novel Gemini vitamin D analog represses the expression of a stem cell marker CD44 in breast cancer. Mol Pharmacol 79:360–367

Sporn MB, Dunlop NM, Newton DL, Smith JM (1976) Prevention of chemical carcinogenesis by vitamin A and its synthetic analogs (retinoids). Fed Proc 35:1332–1338

Srivastava RK, Tang SN, Zhu W, Meeker D, Shankar S (2011) Sulforaphane synergizes with quercetin to inhibit self-renewal capacity of pancreatic cancer stem cells. Front Biosci (Elite Ed) 3:515–528

Strope SA, Andriole GL (2010) Update on chemoprevention for prostate cancer. Curr Opin Urol 20:194–197

Subramaniam D, Ramalingam S, Houchen CW, Anant S (2010) Cancer stem cells: a novel paradigm for cancer prevention and treatment. Mini Rev Med Chem 10:359–371

Syed DN, Suh Y, Afaq F, Mukhtar H (2008) Dietary agents for chemoprevention of prostate cancer. Cancer Lett 265:167–176

Tang SN, Singh C, Nall D, Meeker D, Shankar S, Srivastava RK (2010) The dietary bioflavonoid quercetin synergizes with epigallocathechin gallate (EGCG) to inhibit prostate cancer stem cell characteristics, invasion, migration and epithelial-mesenchymal transition. J Mol Signal 5:14

Tang SN, Fu J, Nall D, Rodova M, Shankar S, Srivastava RK (2012) Inhibition of sonic hedgehog pathway and pluripotency maintaining factors regulate human pancreatic cancer stem cell characteristics. Int J Cancer 131:30–40

Trump DL, Smith DC, Stiff D, Adedoyin A, Day R, Bahnson RR, Hofacker J, Branch RA (1997) A phase II trial of all-trans-retinoic acid in hormone-refractory prostate cancer: a clinical trial with detailed pharmacokinetic analysis. Cancer Chemother Pharmacol 39:349–356

Tu SM, Lin SH (2012) Prostate cancer stem cells. Clin Genitourin Cancer 10:69–76

van der Horst G, van den Hoogen C, Buijs JT, Cheung H, Bloys H, Pelger RC, Lorenzon G, Heckmann B, Feyen J, Pujuguet P et al (2011) Targeting of alpha(v)-integrins in stem/progenitor cells and supportive microenvironment impairs bone metastasis in human prostate cancer. Neoplasia 13:516–525

Vanden Berghe W (2012) Epigenetic impact of dietary polyphenols in cancer chemoprevention: lifelong remodeling of our epigenomes. Pharmacol Res 65:565–576

Venkateswaran V, Klotz LH (2010) Diet and prostate cancer: mechanisms of action and implications for chemoprevention. Nat Rev Urol 7:442–453

Vira D, Basak SK, Veena MS, Wang MB, Batra RK, Srivatsan ES (2012) Cancer stem cells, microRNAs, and therapeutic strategies including natural products. Cancer Metastasis Rev 31(3–4):733–751

Visvader JE, Lindeman GJ (2008) Cancer stem cells in solid tumours: accumulating evidence and unresolved questions. Nat Rev Cancer 8:755–768

Wang Z, Zhang Y, Banerjee S, Li Y, Sarkar FH (2006) Notch-1 down-regulation by curcumin is associated with the inhibition of cell growth and the induction of apoptosis in pancreatic cancer cells. Cancer 106:2503–2513

Wang XB, Zhu L, Huang J, Yin YG, Kong XQ, Rong QF, Shi AW, Cao KJ (2011) Resveratrol-induced augmentation of telomerase activity delays senescence of endothelial progenitor cells. Chin Med J (Engl) 124:4310–4315

Wu M, Shi L, Cimic A, Romero L, Sui G, Lees CJ, Cline JM, Seals DF, Sirintrapun JS, McCoy TP et al (2012) Suppression of Tak1 promotes prostate tumorigenesis. Cancer Res 72:2833–2843

Yi SY, Nan KJ (2008) Tumor-initiating stem cells in liver cancer. Cancer Biol Ther 7:325–330

Yu Y, Kanwar SS, Patel BB, Nautiyal J, Sarkar FH, Majumdar AP (2009) Elimination of colon cancer stem-like cells by the combination of curcumin and FOLFOX. Transl Oncol 2:321–328

Zhang LL, Li L, Wu DP, Fan JH, Li X, Wu KJ, Wang XY, He DL (2008) A novel anti-cancer effect of genistein: reversal of epithelial mesenchymal transition in prostate cancer cells. Acta Pharmacol Sin 29:1060–1068

Zhang L, Li L, Jiao M, Wu D, Wu K, Li X, Zhu G, Yang L, Wang X, Hsieh JT et al (2012) Genistein inhibits the stemness properties of prostate cancer cells through targeting Hedgehog-Gli1 pathway. Cancer Lett 323:48–57

Zhou W, Kallifatidis G, Baumann B, Rausch V, Mattern J, Gladkich J, Giese N, Moldenhauer G, Wirth T, Buchler MW et al (2010) Dietary polyphenol quercetin targets pancreatic cancer stem cells. Int J Oncol 37:551–561

# Chapter 9
# Stem Cell Models for Functional Validation of Prostate Cancer Genes

**Lindsey Ulkus, Min Wu, and Scott D. Cramer**

**Abstract** Prostate cancer is a genomically complex disease in which initiation, progression, and metastasis are regulated by numerous molecular processes including oncogene activation or tumor suppressor inactivation. Understanding the molecular mechanisms that drive prostate tumorigenesis has important clinical implications. Putative oncogenes or tumor suppressors are identified using technologies including SNP arrays, microarrays, and whole genome sequencing, but these targets must then be evaluated in cell and animal models to determine the functional consequences of these genomic alterations. Traditionally, potential prostate cancer genes have been validated with human prostate cancer cell line models (i.e., tissue culture and xenograft systems) or genetically engineered mouse (GEM) models. More recently, stem cell models have been utilized to evaluate candidate cancer genes. Because the normal adult prostate stem cell (PSC) shares many properties with the prostate tumor-initiating cell (TIC) including the capabilities for self-renewal, differentiation, and androgen independence, modeling gene alterations in PSCs may be more appropriate than traditional approaches. PSCs can be maintained in cell culture, genetically manipulated, and characterized using techniques including cell sorting, colony formation assays, and prostasphere assays in vitro and tissue recombination in vivo. A number of prostatic oncogenes and tumor suppressors including MYC, ERG, PTEN, P53, NKX3.1, and TAK1 have been evaluated using stem cell models. Compound genetic alterations have also been studied using

L. Ulkus, Ph.D. • S.D. Cramer, Ph.D. (✉)
Department of Pharmacology, University of Colorado, Anschutz Medical Campus, Aurora, CO 80045, USA

Cancer Biology Training Program, Wake Forest University School of Medicine, Winston-Salem, NC 27145, USA
e-mail: scott.cramer@ucdenver.edu

M. Wu, Ph.D.
Department of Pharmacology, University of Colorado, Anschutz Medical Campus, Aurora, CO 80045, USA

S.D. Cramer (ed.), *Stem Cells and Prostate Cancer*, DOI 10.1007/978-1-4614-6498-3_9, 

PSC models. In this chapter we describe current approaches being used to investigate putative oncogenes and tumor suppressors in the context of the PSC and highlight a few examples of recent studies using stem cell models for target validation. We also discuss the limitations of existing models as well as strategies to improve upon these models for future studies.

## 9.1 Introduction

A major goal of prostate cancer research is to identify the molecular mechanisms that drive the initiation, progression, and metastasis of prostate tumors. Understanding the molecular processes that regulate prostate tumorigenesis can lead to the development of new diagnostic, prognostic, and therapeutic alternatives to improve clinical outcome especially for patients with metastatic disease. However, identification of key regulators of prostate cancer is a challenging task because the prostate cancer genome is extremely complex, and tumor development is driven by a variety of molecular events. Events that can initiate tumorigenesis include amplification, mutation, translocation, or upregulation of an oncogene or deletion, downregulation, or methylation of a tumor suppressor gene. MicroRNAs can also modulate gene expression to promote tumor development (Lei et al. 2006; Lukacs et al. 2010; Shen and Abate-Shen 2010). In a study highlighting the complexity of prostate cancer genetics, Garraway and colleagues observed medians of 3,866 point mutations, 20 nonsynonymous coding mutations, and 90 genomic rearrangements per prostate tumor genome analyzed (Berger et al. 2011). With the advent of technologies including high-powered SNP arrays and whole genomic sequencing, the entire landscape of the prostate cancer genome can be studied with high sensitivity and precision. A number of recent studies have utilized these genome-wide approaches to identify novel regions of genetic alterations that harbor putative oncogenes or tumor suppressors (Berger et al. 2011; Grasso et al. 2012; Ishkanian et al. 2009; Liu et al. 2012; Robbins et al. 2011; Solimini et al. 2012; Taylor et al. 2010). To distinguish genes that are true regulators of prostate tumorigenesis (drivers) from those genes whose alteration does not affect cancer development (passengers), candidate genes must be thoroughly screened using both cell- and animal-based models. Stem cell models have been effectively and efficiently used for evaluation of numerous oncogenes and tumor suppressors including MYC, ERG, PTEN, P53, NKX3.1, and TAK1.

## 9.2 Rationale for Stem Cell Models

Historically, tumorigenesis has been proposed to occur through clonal evolution, a process by which normal cells sporadically mutate and generate progeny that later acquire additional mutations, eventually yielding a heterogenous population of tumor cells in which individual cells equally possess the ability to proliferate and metastasize (Fig. 9.1a) (Greaves and Maley 2012). More recently, another

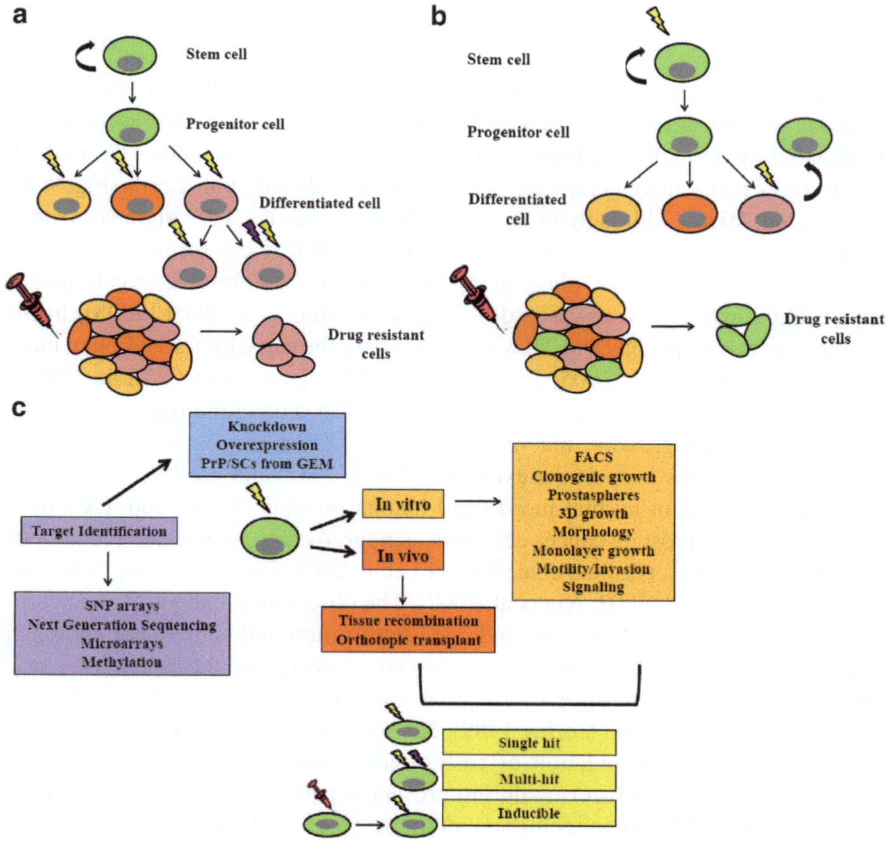


**Fig. 9.1** Advantages of stem cell models for validation of candidate cancer genes. (**a**) Clonal evolution model. In this model tumorigenesis is proposed to occur through the accumulation of mutations or alterations in normal cells (not necessarily stem cells). Daughter cells inherit mutations and can acquire additional alterations, eventually resulting in a heterogenous population of tumor cells in which individual cells equally possess the ability to proliferate and metastasize. Therapeutic intervention may eliminate the bulk of the tumor cell population, but often a small population of drug-resistant tumor cells emerges. These cells may have acquired additional or different mutations than the bulk tumor population. (**b**) Cancer stem cell hypothesis. In this model, tumor development is initiated and sustained by a small population of CSCs or TICs that have properties similar to normal stem cells including the abilities to self-renew, differentiate, and become drug resistant. The CSC may be derived from a normal adult stem cell that has mutated. The CSC may also originate from a more differentiated cell that has mutated and reverted back to a stem-like state. (**c**) Utilization of stem cell models for evaluation of putative oncogenes and tumor suppressors. Candidate cancer genes are identified using technologies including SNP arrays, whole genome sequencing, microarrays, and methylation assays. These targets are then manipulated in PrP/SCs in vitro to suppress or overexpress candidate gene(s). PrP/SCs can also be isolated from GEM mice that already possess a particular genomic alteration (e.g., Pten deletion). Manipulated PrP/SCs are assessed for phenotypic changes in vitro by examining various cellular properties including cell morphology, proliferation rate, migration rate, and signaling ability. Growth can be evaluated in standard monolayer growth assays, clonogenic assays, prostasphere assays, or 3D growth assays. PrP/SCs are characterized in vivo using tissue recombination or orthotopic models. PrP/SC models can be adapted to examine the effects of multiple genetic alterations on prostate tumorigenesis (multi-hit/compound models), and this can be done using regulated genetics (drug-inducible models)

hypothesis for tumorigenesis has emerged; the cancer stem cell (CSC) hypothesis proposes a hierarchy for cancer development in which tumors are initiated and propagated by a small population of CSCs or tumor-initiating cells (TIC) which may be more resistant to therapeutic agents than more differentiated cells (Fig. 9.1b) (Wicha et al. 2006). Although it remains unclear whether the TIC originates from a mutated adult stem cell or a more differentiated cell that has reverted to a more stem-like state, the TIC does share many properties of the normal stem cell including the abilities to self-renew and differentiate (Collins et al. 2001, 2005). In the prostate, normal stem cells and TICs may also share the capacity for androgen independence (Collins et al. 2005; English et al. 1987; Evans and Chandler 1987). Therefore, the normal prostate adult stem cell (PSC) is a natural model for studying tumor initiation, progression, and metastasis.

Numerous studies support the existence of both PSCs and prostate TICs. PSCs have been detected in normal mouse and human prostatic tissue, and TICs have been identified in prostate cancer cell lines, xenografts, and mouse and human primary prostate tumors (Collins et al. 2001; Li et al. 2008; Patrawala et al. 2006; Shi et al. 2007; Tran et al. 2002; Xin et al. 2007). The observation that basal and neuroendocrine cells survive following androgen deprivation and subsequent androgen restoration while most luminal cells undergo apoptosis during the regression phase suggests that PSCs may reside in the basal layer (English et al. 1987; Evans and Chandler 1987; Isaacs and Coffey 1989). More recently, Shen and colleagues discovered that a small population of castration-resistant luminal cells expressing Nkx3.1 (CARNs) can self-renew during prostate regeneration, providing evidence for stem cells in the luminal cell population (Wang et al. 2009). Since then, other lineage-tracing experiments have provided evidence that both basal and luminal cells contain populations of PSCs (Choi et al. 2012; Liu et al. 2011). Although the origin of the PSC continues to be debated, it is clear that identification, characterization, and manipulation of the PSC will provide a better understanding of tumor development, resistance to treatment, recurrence, and metastasis.

## 9.3 Advantages of Stem Cell Models over Traditional Strategies for Validation of Cancer Genes

The many similarities between PSCs and TICs (described in Sect. 9.2) make the PSC a natural model for functional validation of prostate oncogenes and tumor suppressors. With the discovery of PSC and TIC biomarkers and the optimization of assays for characterization of stem cells, utilization of PSCs in functional studies is now possible. Traditionally, the approach to validating putative oncogenes and tumor suppressors has been to manipulate gene expression in human prostate cancer cell lines, determine if there are differences in cellular properties in vitro, and then subcutaneously graft cells into nude mice to assess effects on tumor size in vivo. Although experiments performed with prostate cancer cell lines are fairly inexpensive, easy, and fast, these cells may not accurately reflect the true genotypic and phenotypic characteristics of prostate tumors. In fact, the three most commonly used prostate cancer cell lines DU145, PC3, and LNCaP were established from brain, bone, and lymph node metastases, respectively, and harbor numerous and diverse genomic alterations (Liu et al. 2008; van Bokhoven et al. 2003). When a gene is manipulated in a cell population that is already highly altered, the conclusions that can be drawn about that gene's role in tumor promotion or initiation must be tempered by uncontrollable interactions with other genetic alterations. In support of this concept, we have observed that gene suppression can have different or even opposite effects in LNCaP, DU145, and PC3 cells which is likely due to the vast genetic diversity among these cancer cell lines (Ulkus and Cramer, unpublished observations).

If genetic manipulation causes an abnormal phenotype in the human cell line/xenograft system, then there is often rationale for creation of a genetically engineered or transgenic mouse (GEM) model. GEM models introduce genetic alterations into the genomes of embryonic stem (ES) cells allowing for germline incorporation of mutations, deletions, duplications, or other modifications. GEM models are useful for studying tumorigenesis because these mice can develop multifocal prostate cancer that can recapitulate different aspects of the human disease. Types of GEM models that have been used in prostate studies include:

1. First generation transgenic models that utilize prostate tissue-specific promoters such as the rat probasin (PB) gene to drive expression of viral oncogenes (e.g., TRAMP, LADY)
2. Transgenic models that conditionally express nonviral oncogenes (e.g., c-Myc, Akt, FGFR1, Braf)
3. Second generation GEM models that introduce whole body gene knockout by homologous recombination in ES cells (e.g., Pten, Rb)
4. Third generation conditional GEM models that induce genetic alterations in a cell or tissue-specific manner most commonly using Cre-loxP technology (e.g., PB-Cre; $Pten^{L/L}$)
5. Regulated GEM models that utilize drug-inducible promoters to introduce genomic alterations at specific time points in development (e.g., PSA-Cre-$ER^{T2}$; $Pten^{L/L}$ or K14-Cre$ER^{T2}$; $Pten^{L/L}$)
6. Compound GEM models that introduce two or more genomic alterations simultaneously (e.g., PB-Cre; $Pten^{L/L}$;$p53^{L/L}$ or PB-Cre; $Pten^{L/L}$; $Kras^{G12D/W}$)

Overall, GEM models have been highly successful for validation of putative oncogenes and tumor suppressors in the prostate, but the major limitation of these models is that they cannot fully recapitulate human tumorigenesis. Advanced stage human prostate cancers commonly metastasize to the bone, but currently no GEM model can consistently form osteoblastic bone metastases. Additionally, some GEM models display phenotypes inconsistent with human pathology. In the TRAMP model, for example, tumors often originate in neuroendocrine cells, a phenotype rarely observed in human prostate cancer (Chiaverotti et al. 2008). In a probasin-cH-RAS-G12V model, mice display low-grade prostatic intraepithelial neoplasia (PIN) as early as 3 months of age but also develop intestinal metaplasia (Scherl et al. 2004). Also, according to the CSC hypothesis, genetic alterations occur in the TIC which is presumably an adult prostate cell with stem cell-like properties. With the exception of the more recently developed inducible promoter models, which may introduce alterations into the adult prostate cell, all other GEM models target the ES cell. Since alteration of the ES cell impacts embryonic and postnatal development, this strategy may not accurately reflect the initiation and progression of prostate cancer. Even in models that utilize the tamoxifen-regulated inducible Cre-ER$^{T2}$ system, the only tissue-specific promoters available include probasin, PSA, and NKX3.1. These promoters are in reality more cell specific rather than tissue specific because the proteins are predominantly expressed in the (more differentiated) luminal cells of the prostate (Luchman et al. 2008; Ratnacaram et al. 2008; Wang et al. 2009). Other cell-specific promoters utilized in the Cre-ER$^{T2}$ model include K8, a luminal cell-specific cytokeratin and K14, a cytokeratin expressed in the basal cell (Choi et al. 2012). However, a prostate epithelial stem cell-specific promoter has not been identified.

Another issue that can make interpreting results from GEM models difficult is that if the same transgene is introduced into mice originating from different genetic strains, the resulting phenotypes may not be consistent. To study the effects of oncogenic ERG on prostate tumorigenesis, multiple groups have generated transgenic mice that overexpress ERG or TMPRSS2:ERG under the control of the probasin promoter. However, the results from studies using PB-ERG mice have been conflicting. Two reports demonstrate that overexpression of ERG leads to the development of PIN in FVB and 129/SV mice (Klezovitch et al. 2008; Tomlins et al. 2008). However, three studies show that transgenic TMPRSS2:ERG or ERG mice display minor phenotypic changes that do not progress to PIN in FVB and B6J mice (Carver et al. 2009; Casey et al. 2012; King et al. 2009). Contrasting findings may be attributable to minor differences in ERG vector design and integration into the mouse ES cell. However, these findings may also be a result of differences in mouse genetic backgrounds. Finally, GEM models are very time consuming and costly to generate. These models require complex vector design and breeding schemes which can take years to validate and study.

PSC models have many advantages over standard cancer cell line and GEM models. Critically, if the CSC hypothesis holds true, then genetic manipulation of the PSC may better recapitulate tumor development and progression compared to modeling tumorigenesis in a more differentiated cell type. Genetic alteration of an oncogene or tumor suppressor in the PSC is predicted to have visible phenotypic consequences such as altered cellular proliferation, motility, morphology, signaling, and/or abnormal glandular development. Standard biological assays can be used to qualify and quantify these phenotypes. Furthermore, from a practical perspective, stem cell models are less expensive, less time consuming, and less technically challenging to work with compared to GEM models. Isolation, manipulation, and both in vitro and in vivo characterization of PSCs can be completed in a few months. Working with PSCs also affords the researcher a high degree of flexibility at each step in the process of *isolation*, *culture*, *manipulation*, and *characterization*. PSCs can be *isolated* from prostates originating from various mouse backgrounds (Barclay and Cramer 2005) including:

1. Different mouse strains (e.g., C57BL6 or FVB)
2. Mice of varying ages (e.g., embryonic, adolescent, or adult)

3. Mice of varying genetic backgrounds including single and compound GEM
4. Mice with inducible or regulated gene expression

Under optimized tissue culture conditions, PSCs can be *cultured* over many passages while retaining their stem cell properties (Barclay et al. 2008). PSCs can be *manipulated* to alter gene expression using common strategies for gene overexpression or suppression including transient transfection and transduction with lentiviral, adenoviral, or retroviral vectors. PSCs can be infected with drug-inducible vectors to study temporal effects of changes in gene expression or transduced with multiple vectors to easily create compound models of gene alterations. Also, different PSC populations can be purified and studied separately using FACS technology (discussed in Chap. 1). PSCs can be *characterized* using in vitro models including prostasphere/three-dimensional (3D) assays and colony formation assays (described in detail in Chap. 2) as well as in vivo systems including the tissue recombination model and the prostate orthotopic model (Fig. 9.1c). In the next two sections of this chapter, we highlight the importance of PSC culture and tissue recombination systems.

Of course, even PSC models have limitations because none of these systems can exactly recapitulate the adult prostatic stem cell in its natural state. For instance, in vitro assays maintain stem cells in an artificial state in the absence of the prostatic microenvironment. Tissue culture conditions put selective pressure on cells and they can adapt by acquiring genetic mutations or other changes so "stem-like cells" may be sustainable in vitro, but stem cells may not. Likewise, the PSC is not manipulated in its native state in in vivo assays because genetic alterations are introduced into the PSC before normal prostatic glandular development occurs. From this point on, we refer to the cells described in PSC models as prostate progenitor/stem cells (PrP/SCs) to reflect the fact that these cells are not native PSCs but do maintain stem cell characteristics. Because no one stem cell model can simulate the true nature of the prostate stem cell, it is important to evaluate candidate prostate cancer genes using multiple assays. The in vitro and in vivo stem cell models mentioned above can be used in combination to provide a more complete assessment of prostate cancer genes and their roles in tumorigenesis.

## 9.4 PSC Culture Models

In the 1980s, numerous research groups began to experiment with various culture conditions and mitogens required for sustained growth of normal and malignant human and rodent epithelial cells from various tissues including the breast and prostate (Imagawa et al. 1982; Kubota et al. 1981; McKeehan et al. 1982; Peehl and Stamey 1984, 1986). Nandi and colleagues first utilized a collagen gel matrix system to improve in vitro propagation of mouse mammary epithelial cells, and this method has since been adapted to culture of mouse prostate epithelial cells (MPECs) (Kusama et al. 1989; Yang et al. 1980). We have developed a modified culture method in which epithelial organoids are plated on collagen-coated dishes and remain on collagen until MPEC outgrowths survive crisis at which time they are transferred to plastic dishes and serially passaged (Barclay and Cramer 2005). We have isolated MPECs from various genotypes including wild-type B1/6;129/SVEV, $Rb^{L/L}$ FVB129-Rb1tm2Bm, $Ink4a^{-/-}$ B6;129-Cdkn2atm1Rdp, $VDR^{-/-}$ B6;CD-1; $Pten^{L/L}$ C57BL/6, and $Map3k7(Tak1)^{L/L}$ C57BL/6 mice. During in vitro characterization of wildtype B1/6;129/SVEV MPECs, hereafter called WFU3 cells, we first observed that these cells possess PrP/SCs properties. When WFU3 cells were weaned from serum (1% to 0% fetal bovine serum (FBS)) in monolayer cultures, they formed patches of tightly clumped cells resembling spheroids. Spheres collected from monolayer cultures and replated in 3D collagen matrices formed large, branched ductal-like structures (Barclay and Cramer 2005). To confirm the existence of PrP/SCs within the WFU3 cell population, we tested the differentiation capability of WFU3 cells using the tissue recombination system (see Sect. 9.5 for a detailed description of this model). WFU3 tissue recombinants regenerated normal prostatic ductal structures with a p63+ basal layer and an AR+ lumen that produced prostatic secretory products. WFU3 PrP/SCs also express high levels of Sca1 and Cd49f and possess the ability to self-renew upon serial transplantation of tissue recombinants in vivo (Barclay et al. 2008). Because WFU3 cells maintain their PrP/SC characteristics over long-term culture, they are an invaluable PSC model and can be genetically manipulated to determine effects of genomic alteration(s) on numerous biological processes including normal development and tumorigenesis. To date, WFU3 cells have been manipulated to alter expression of numerous genes including Pten, IL1-α, Chd1, and Map3k7/Tak1 (Axanova et al. 2010; Liu et al. 2012; Maund et al. 2011; Wu et al. 2012).

Interestingly, mixed populations of MPECs isolated by our culture technique from various genetic backgrounds differ in their proportions of PrP/SCs and thus their ability to generate prostatic ductal structures in tissue recombinants. We found that tissue recombinants from bulk populations of $Rb^{L/L}$ and $Pten^{L/L}$ MPECs generated prostatic ductal structures while $Tak1^{L/L}$ MPECs did not (Barclay et al. 2008 and unpublished observations). Flow cytometry of these populations of MPECs stained with Sca1 and Cd49f antibodies revealed that 30.6% of $Rb^{L/L}$, 12.5% of $Pten^{L/L}$, and 2.8% of $Tak1^{L/L}$ MPECs cells had a $Sca1^{+}Cd49f^{hi}$ phenotype. The low percentage of PrP/SCs within the total $Tak1^{L/L}$ population likely explains the inability of $Tak1^{L/L}$ cells to form ductal structures in vivo. Single cells were plated from sorted $Sca1^{+}Cd49f^{hi}$ $Tak1^{L/L}$ cells and clonal populations of PrP/SCs were established. Unlike the bulk $Tak1^{L/L}$ cells which grew in a single layer on plastic dishes, numerous $Sca1^{+}Cd49f^{hi}$ $Tak1^{L/L}$ clones formed spheroids under monolayer culture conditions even in the presence of serum and after serial passaging (Fig. 9.2b). The same clones generated large, branched structures in 3D collagen matrices (Fig. 9.2c). Although the majority of 3D structures were branched, we observed the presence of some rounded structures mixed in among the branched structures (Fig. 9.2c, *left*) (Romero and Cramer, unpublished). Variation in 3D structure may be indicative of PrP/SCs at different stages of differentiation. Further characterization by staining individual structures with markers of prostatic stemness and differentiation can provide more insight into

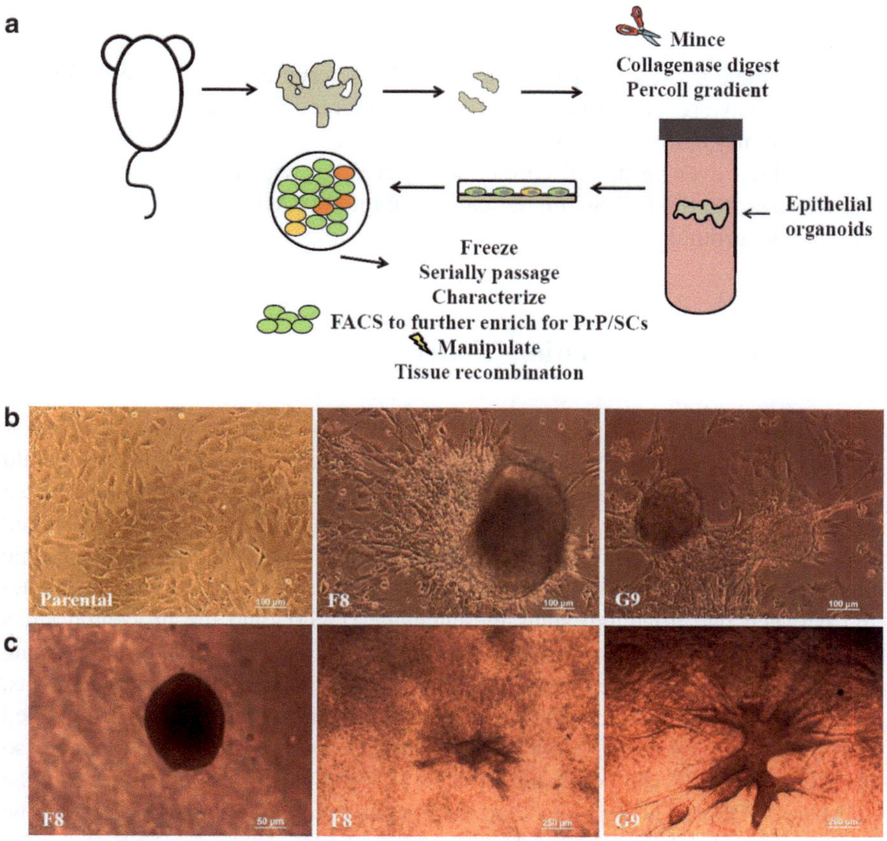


**Fig. 9.2** PrP/SC culture model. (**a**) Strategy for isolation and culture of PrP/SCs. Prostates are removed from 6- to 8-week-old mice and the different prostatic lobes (anterior, dorsal, lateral, and ventral) are isolated. Lobes from age and genetically matched mice can be pooled. Prostatic tissues are then minced, collagenase digested, and transferred to a Percoll gradient to separate epithelial cells from stromal cells and debris using gradient centrifugation. Epithelial organoids are collected and plated on collagen-coated plastic dishes in DMEM/F12 medium supplemented with FBS, bovine serum albumin (BSA), cholera toxin, epidermal growth factor (EGF), bovine pituitary extract (BPE), insulin, transferrin, Vitamin E, trace elements, and gentamicin (Barclay and Cramer 2005). Once cells survive crisis, they can be cultured on regular tissue culture plates. Cells can then be frozen, serially passaged, characterized in vitro, genetically manipulated, and used in tissue recombination experiments in vivo. FACS can also be used to enrich for PrP/SCs before characterization, manipulation, or tissue recombination. (**b**) Enrichment for $Tak1^{L/L}$ PrP/SCs and growth in monolayer culture. $Tak1^{L/L}$ MPECs were isolated from the anterior lobes of 7-week-old mice and cultured as illustrated in Fig. 9.2a. Parental (unsorted) $Tak1^{L/L}$ MPECs display a typical cobblestone-like pattern when grown in monolayer culture conditions (*left*). $Tak1^{L/L}$ MPECs can be enriched for PrP/SCs using FACS. $Tak1^{L/L}$ MPECs were co-stained with anti-Sca1-APC and anti-Cd49f -FITC antibodies and FACS sorted. Single $Sca1^{+}Cd49f^{hi}$ PrP/SCs were sorted directly into individual wells of a 96-well plate. Clonal populations were established, expanded, and serially passaged. $Tak1^{L/L}$ PrP/SC clonal cell lines F8 and G9 form spheroid-like structures in monolayer culture even in the presence of serum (*middle* and *right*). (**c**) Characterization of $Tak1^{L/L}$ PrP/SCs in 3D culture. $1 \times 10^4$ cells in a collagen matrix mixture were plated in wells of a 24-well collagen-coated plate and maintained in culture for 21 days. Media was refreshed every other day. $Tak1^{L/L}$ PrP/SC clones F8 and G9 generated large, branched structures in 3D culture (*middle* and *right*). Clone F8 also formed some small, rounded structures when grown in collagen suggesting that not all cells within this population have retained their PrP/SC characteristics (*left*)

this phenomenon. Based on their in vitro characteristics, $Sca1^{+}Cd49f^{hi}$ $Tak1^{L/L}$ cells likely have regenerative capability in tissue recombination and this prediction is currently being tested. We have demonstrated that PrP/SCs can be enriched and sustained even in mixed epithelial populations with low numbers of stem cells. Therefore, the PrP/SC model can be adapted to MPECs isolated from any genetic background.

## 9.5 Tissue Recombination Model

The mouse prostate tissue recombination model, also called the prostate regeneration model or the prostate reconstitution model, was originally developed by Gerald Cuhna in the 1970s (Cunha 1972a, b) and has since been modified by others (see Fig. 9.3a, model adapted by Cramer lab). Cunha's initial experiments were designed to study the interaction between the mesenchyme and epithelium during embryonic development of the mouse prostate, and he demonstrated that a reciprocal interaction between the urogenital sinus mesenchyme (UGM) and the urogenital sinus epithelium (UGE) is critical for (1) development and differentiation of the prostate epithelium and (2) differentiation of the mesenchyme into stromal components. In classical recombination experiments, the urogenital sinus (UGS) is removed from a mouse or rat during embryogenesis prior to prostatic bud development (so that UGM and UGE can be separated cleanly). Following enzymatic digestion, the UGM and UGE can be separated and manipulated individually. UGM can then be recombined with UGE ex vivo and grafted under the renal capsule of a host mouse. After several weeks or months, grafts are harvested and the histology is evaluated (see Fig. 9.3b, macroscopic view of grafts after kidney removal). Cunha discovered that grafts containing both UGM and UGE generate fully differentiated prostatic structures demonstrating that the UGS possesses all cell types including stem/progenitor cells required for normal prostate development (Cunha 1972a).

Since these pioneering experiments, the tissue recombination model has been applied to study prostatic tissues from normal adult mice, GEM, prostate cancer cell lines, and primary human prostate samples to evaluate the tumorigenic potential of the prostate epithelium. Additionally, gene expression has been manipulated in MPECs, PrP/SCs, and stromal cells in vitro using retroviruses or lentiviruses, and the effects of various genetic alterations on normal prostatic development and differentiation have been assessed using the tissue recombination system in vivo. Key experiments by Thompson and colleagues in the late 1980s first demonstrated the utility of tissue recombination models for investigating the functional significance of genetic alterations in prostate tumorigenesis. In the model used by Thompson and colleagues, the intact UGS was retrovirally infected with activated Ras and/or Myc, two oncogenes commonly overexpressed in numerous cancer types. Briefly, UGS cells were isolated from E16- to E17-day-old embryos, dissociated to form single-cell suspensions, and inoculated with retrovirus for 2 h. Retrovirally infected UGS was then implanted under the renal capsule. In grafts of UGS infected with

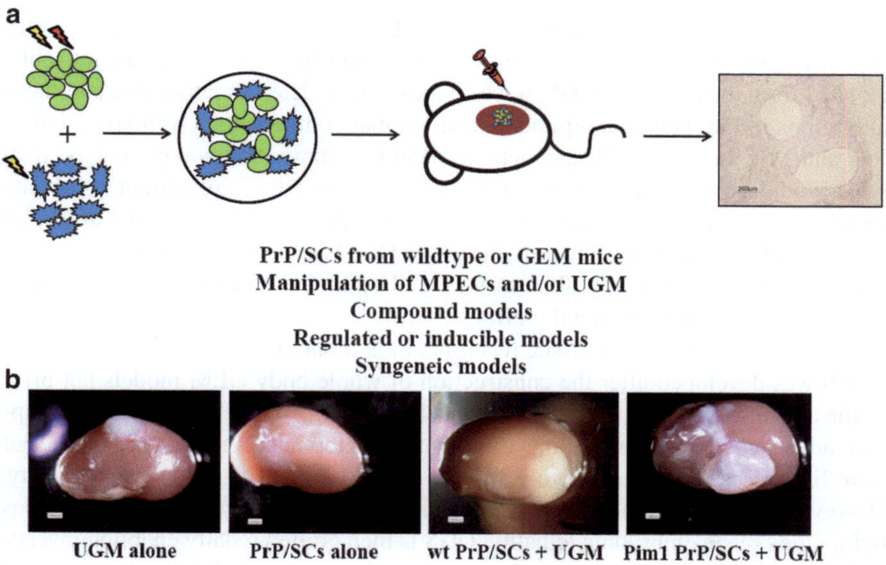


**Fig. 9.3** Tissue recombination model. (**a**) Strategy for tissue recombination and renal grafting. $1 \times 10^5$ cells PrP/SCs are combined with $2.5 \times 10^5$ embryonic day 18 rat UGM cells and resuspended in collagen. Recombinants are plated as collagen buttons and incubated overnight at 37 °C before implantation under the renal capsule of male mice. After 6–12 weeks, kidneys are removed, grafts are formalin fixed, and histology is assessed using immunohistochemistry methods. Histology of a normal ductal structure from a wildtype PrP/SC recombinant is shown here. The tissue recombination model can be adapted to evaluate alterations in epithelial and/or mesenchymal cells (indicated by *yellow bolts*) or multiple alterations (*red bolt*). Alterations can also be introduced into PrP/SCs post-grafting using drug-inducible models. (**b**) Renal grafts of Pim1-overexpressing PrP/SCs are larger than PrP/SC wt grafts. PrP/SC wt or PrP/SC Pim1 recombinants were grown under the renal capsules of immunocompromised mice for 12 weeks. Mice were then sacrificed and kidneys were removed. Grafts of UGM alone or PrP/SCs alone form very small grafts while grafts from wt PrP/SCs + UGM and Pim1-overexpressing PrP/SCs + UGM tissue recombinants form large, macroscopic grafts. Although Pim1-overexpressing PrP/SCs + UGM tissue recombinants appear to form the largest grafts, histological analysis is the only conclusive way to determine if Pim1 overexpression effects prostatic development in this model (not shown)

low-titer Ras retrovirus, normal prostatic structures formed, while in high-titer Ras-infected grafts, focal dysplasia, hyperproliferative stroma, and increased angiogenesis were observed relative to grafts with control retrovirus. Myc-infected UGS grafts demonstrated a mild hyperplasia mostly confined to epithelial regions. UGS co-infected with Ras and Myc formed rapidly growing carcinomas even at low viral titers suggesting that co-activation of Ras and Myc is sufficient to induce prostate tumorigenesis (Thompson et al. 1989). Subsequently, GEM Ras models have supported these findings (Scherl et al. 2004). Later, UGS from $p53^{+/-}$ or $p53^{-/-}$ mice was isolated and infected with the Ras-Myc retrovirus. One hundred percent of Ras-Myc retrovirally infected $p53^{+/-}$ and $p53^{-/-}$ UGS grafts formed carcinomas and 95% of mice had metastases (Thompson et al. 1995). These experiments provided the first evidence that multiple genetic alterations can be modeled using the tissue recombination system. However, because activated Ras and Myc were introduced into cells isolated from the complete UGS, it was impossible to determine the effects of these alterations in the individual epithelial and stromal components. Overexpression of Ras and Myc in total UGS may actually lead to a more dramatic phenotype than what would be observed if these were overactive in either epithelium or stroma alone. Since adult MPECs contain PrP/SC populations that possess the ability to form ductal structures when recombined with UGM (as described in Sect. 9.4), it is now possible to genetically alter stromal and epithelial compartments separately before recombination and renal grafting.

A useful variation of the tissue recombination model is the tissue rescue system which was developed after the construction of whole body GEM models but prior to the advent of tissue-specific GEM models. Whole body GEM models have rapidly advanced our understanding of normal development and tumorigenesis but have limited use in cases where complete gene loss leads to embryonic lethality. However, using the tissue rescue approach, UGS can be isolated from embryos before they reach embryonic lethality. UGS is then grafted renally enabling investigation of the effects of gene deletion on normal prostatic ductal development and tumor initiation. Tissue rescue has been used to characterize UGS from $Rb^{-/-}$ and $p57^{-/-}$ mice. RB and P57 are often lost or downregulated in prostate cancer and nullizygous mice of either genetic background display early lethality (Clarke et al. 1992; Jacks et al. 1992; Jin et al. 2008; Lee et al. 1992). Cunha and colleagues utilized the tissue rescue approach to test the functional significance of complete Rb loss in prostate cancer development. Pelvic visceral rudiments were isolated from day E12 embryos and grafted under the renal capsule of athymic male mice for 4 weeks to generate $Rb^{+/+}$ or $Rb^{-/-}$ prostatic tissue containing characteristic ductal structures. Prostatic epithelial ducts were then removed from the renal capsule, recombined with rat UGM, sub-renally grafted into new athymic hosts along with implantation of an empty capsule or a capsule containing estradiol and testosterone (a drug combination known to induce carcinogenesis in rats) for 5–8 weeks. $Rb^{+/+}$ and $Rb^{-/-}$ grafts without hormonal stimulation (empty capsule) formed mostly normal structures with some hyperplastic foci observed. $Rb^{+/+}$ grafts with hormone treatment generated normal and hyperplastic structures. $Rb^{-/-}$ grafts plus hormone formed normal, hyperplastic, atypical hyperplastic, and carcinomatous structures suggesting that loss of Rb makes the prostate more susceptible to prostate cancer development (Wang et al. 2000). The development of a prostate-specific conditional Rb knockout mouse (Pb-Cre) not only supported the findings of Cunha and colleagues but also demonstrated a direct association between Rb loss and prostate carcinogenesis. Loss of one or both copies of Rb resulted in prostatic hyperplasia in 18-week-old mice but never progressed to carcinoma, suggesting that Rb may be a haploinsufficient prostate tumor suppressor that can initiate prostate cancer but that additional genetic events are required for progression to carcinoma (Maddison et al. 2004). The tumor suppressive ability of p57 was also verified in tissue rescue experiments using UGS isolated from day E15.5 to E18.5 $p57^{-/-}$ mouse embryos. By 2 and 4 months after grafting, $p57^{-/-}$ UGS cells displayed hyperplasia and increased proliferation relative to wildtype UGS grafts. After 6 months, $p57^{-/-}$ grafts displayed

PIN and carcinoma (Jin et al. 2008). With the advent of the tissue-specific knockout mouse, the tissue rescue approach became outdated, but it served as an initial proof of principle that nullizygous genotypes can be studied using PrP/SCs.

A major advantage to using the tissue recombination model is its versatility. Not only can stromal and epithelial cells from various mouse, rat, and human origins be studied in this system, but the grafts can be placed into hosts of various backgrounds including syngeneic mice (Fig. 9.3a). Unlike with GEM models, complex breeding schemes are not necessary. A number of studies have utilized tissue recombination to validate putative oncogenes and tumor suppressors including Ras, Myc, Pten, Tak1, and Erg by manipulating gene expression in MPECs or PrP/SCs in vitro and studying the phenotype in vivo (see Table 9.1). Furthermore, specific stem cell populations can be purified using markers like Cd49f and Sca1 (described in detail elsewhere in this book) before being grafted renally. Also, tissue recombination can be performed using MPECs or PrP/SCs isolated from single or compound GEM mutants. Although most studies have focused on altering gene expression in epithelial cells, UGM or normal stroma can also be manipulated. Overexpression of FGF10 in UGM causes formation of carcinomas when recombined with normal MPECs or PrP/SCs in renal grafts (Lawson et al. 2010; Memarzadeh et al. 2007).

Although the tissue recombination model is used successfully by many researchers, technical expertise in microdissection and microsurgery is required for separation of UGE from UGM and implantation of grafts under the renal capsule. The procedure typically requires consecutive days of intensive preparation to isolate UGM, recombine with epithelial cells, and perform the grafting surgery. However, recent advances that have improved this process include optimization of culture conditions for expanding UGM (Goldstein et al. 2011) and establishment of UGM cell lines (Shaw et al. 2006). These techniques allow UGM to be grown in cell culture, passaged, and frozen before recombination with epithelial cells. Another limitation of the tissue recombination model is that genetic alterations are typically introduced into UGM or PrP/SCs prior to prostatic development under the renal capsule when in actuality genomic lesions form in the fully developed adult prostate. To better recapitulate tumor development, PrP/SCs with regulatable genetics can be introduced into the tissue recombination system to allow prostate structures to form in vivo before genomic alterations are induced (discussed in Sect. 9.7).

## 9.6 Utilization of Stem Cell Models to Validate Candidate Cancer Genes

A growing number of studies have used stem cell models to evaluate oncogenes and tumor suppressors. Table 9.1 contains a summary of genes that have been investigated using PSC models, the methods used, and conclusions made. In this section, we highlight two examples of how stem cell models have provided valuable insights into the roles of tumor suppressors and oncogenes in promotion of tumorigenesis.

**TAK1**, TGF-β activated kinase-1, encoded by the *MAP3K7* gene on chromosome 6q15 is a serine/threonine kinase that regulates numerous cellular processes

**Table 9.1** Summary of stem cell models used to validate candidate oncogenes and tumor suppressors

| Gene(s) | Method(s) | Outcome(s) | Reference |
|---|---|---|---|
| Rb | Tissue rescue of $Rb^{-/-}$ UGS, Tissue recombination | ↓Rb→hyperplasia and carcinoma | Wang et al. (2000) |
| p57 | Tissue rescue of $p57^{-/-}$ UGS | ↓p57→PIN and carcinoma | Jin et al. (2008) |
| TGFβ1 | Overexpression in UGS, Tissue recombination | ↑TGFβ1→basal cell hyperplasia, stromal abnormalities | Timme et al. (1996) |
| FGF10 | Overexpression in UGM, Tissue recombination with FACS-enriched PrP/SCs | ↑FGF10→carcinoma in $Lin^{-}Sca1^{+}Cd49f^{hi}$ PrP/SC population; $Lin^{-}Sca1^{-}Cd49f^{lo}$ cells do not generate ductal structures | Lawson et al. (2010) |
| Pten | MPECs isolated from wildtype and Pten null mice; FACS sorting for Sca1 | ↓Pten→↑percentage of $Sca1^{+}$ cells | Wang et al. (2006) |
| Pten | MPECs isolated from control and Pten null mice; FACS-enriched $Lin^{-}Sca1^{+}Cd49f^{hi}$ PrP/SCs, Prostaspheres, Tissue recombination | ↓Pten→adenocarcinoma in $Lin^{-}Sca1^{+}Cd49f^{hi}$ PrP/SC grafts | Mulholland et al. (2009) |
| PTEN | Knockdown of PTEN in DU145 cells, Prostaspheres | ↓PTEN→↑self-renewal ability | Dubrovska et al. (2009) |
| AKT | Overexpress Akt in PrP/SCs sorted for Sca1, Tissue recombination | ↑Akt→PIN in $Sca1^{+}$ grafts, mostly normal structures in $Sca1^{-}$ grafts | Xin et al. (2005) |
| TMPRSS2:ERG | MPECs isolated from wildtype and TMPRSS2: ERG transgenic mice, FACS enrichment for PrP/SCs, Prostaspheres | ↑TMPRSS2:ERG→↑ self-renewal ability in $Sca1^{+}Epcam^{+}$ population | Casey et al. (2012) |
| ETV1 | Overexpression in MPECs, Tissue recombination | ↑ETV1→hyperplasia | Zong et al. (2009) |
| Tak1 | Knockdown in PrP/SCs, Tissue recombination | ↓TAK1→PIN and carcinoma | Wu et al. (2012) |
| Pim1 | Overexpression in PrP/SCs, Tissue recombination | ↑PIM1→high grade PIN | Chen and Cramer, unpublished |
| Myc | FACS-enriched $CD133^{hi}$ immortalized primary tumor cells treated with antisense c-myc, Prostaspheres, intrathoracic injection in vivo | Antisense c-myc→↓colony-forming ability and growth, ↓ability of $CD133^{hi}$ prostaspheres to form tumors in vivo | Goodyear et al. (2009) |
| Ras+Myc | Overexpression in UGS, Tissue recombination | ↑Ras→dysplasia; ↑Myc→mild hyperplasia; ↑Ras+↑Myc→carcinoma | Thompson et al. (1989) |

| | | | |
|---|---|---|---|
| Ras+Myc+p53 | Overexpression in $p53^{+/+}$ or $p53^{-/-}$ UGS, Tissue recombination | ↑Ras+↑Myc+↓p53→carcinoma and metastasis | Thompson et al. (1995) |
| FGF10+AKT | FGF10 overexpression in UGM, expression of activated AKT in MPECs, Tissue recombination | ↑FGF10→carcinoma (well differentiated); ↑AKT→PIN; ↑FGF10+↑AKT→high-grade carcinoma (poorly differentiated) | Memarzadeh et al. (2007) |
| ERG+AKT | Overexpression of ERG and AKT in MPECs, Tissue recombination | ↑ERG→hyperplasia, focal PIN; ↑ERG+↑AKT→carcinoma | Zong et al. (2009) |
| ERG+Pten | Overexpression of ERG and knockdown of Pten in MPECs, Tissue recombination | ↑ERG→hyperplasia, focal PIN; ↑ERG+↓Pten→carcinoma | Zong et al. (2009) |
| ERG+AR | Overexpression of ERG and AR in MPECs, Tissue recombination | ↑ERG→hyperplasia, focal PIN; ↑AR→↓number of prostatic tubules formed ↑ERG+↑AR→ invasive carcinoma | Zong et al. (2009) |
| AKT+AR | Overactivation of Akt and/or AR in sorted basal or luminal MPEC populations, Tissue recombination | ↑Akt+↑Erg→PIN in basal cell grafts; ↑AR→occasional PIN in basal cell grafts; ↑Akt+↑AR→high-grade carcinoma in basal cell grafts; luminal cell grafts do not form prostatic structures under any condition | Lawson et al. (2010) |
| AKT+ERG+AR | Sort primary human prostate epithelial cells into basal and luminal populations, Overexpress Akt and Erg or Akt and Erg and AR in sorted populations, Tissue recombination | ↑Akt+↑Erg→PIN in basal cell grafts and no prostatic structure formation in luminal cell grafts; ↑Akt+↑Erg+AR→invasive carcinoma in basal cell grafts | Goldstein et al. (2010) |
| p53+Pten | Widtype and $p53^{-/-}Pten^{-/-}$ MPECs; Prostaspheres, Colony formation assay, Orthotopic transplant | $p53^{-/-}Pten^{-/-}$ MPECs have ↑self-renewal ability and from large, more irregular spheres, form tumors in vivo | Abou-Kheir et al. (2010) |
| p53+Pten | $Rosa26\text{-}Cre^{ER}p53^{L/L}Pten^{L/L}$ MPECs; Prostaspheres, Orthotopic transplant | Tamoxifen-induced co-deletion of p53 and Pten in prostaspheres leads to aggressive tumors, metastasis, and castration-resistance in vivo | Abou-Kheir et al. (2011) |
| Pten+KRAS | PrP/SCs isolated from $PB\text{-}Cre^{+}Pten^{L/L}Kras^{G12D/W}$ mice, Prostaspheres | ↑$Lin^{-}Sca1^{+}Cd49f^{hi}$ PrP/SC population, ↑sphere-forming ability in $PB\text{-}Cre^{+}Pten^{L/L}Kras^{G12D/W}$ cells relative to $PB\text{-}Cre^{+}Pten^{L/L}$ cells | Mulholland et al. (2012) |

(continued)

**Table 9.1** (continued)

| Gene(s) | Method(s) | Outcome(s) | Reference |
|---|---|---|---|
| Pten + Nkx3.1 | Forced expression of Nkx3.1 in wildtype and Pten null MPECs; Tissue recombination | ↓Pten→hyperplasia/PIN; ↓Pten and ↑Nkx3.1→prevents cancer initiation (normal ductal phenotype maintained) | Lei et al. (2006) |
| Bmi-1 + Pten | Knockdown of Bmi-1 in wildtype and $Pten^{L/L}$ MPECs, Prostaspheres, Tissue recombination | ↓Bmi-1→↓sphere-forming ability, ↓self-renewal, ↓prostatic tubule formation in vivo; ↓Bmi-1 and delete Pten→ ↓tumor size relative to single-hit Pten null model | Lukacs et al. (2010) |
| Bmi-1 + FGF10 | Knockdown of Bmi-1 in MPECs and FGF10 overexpression in UGM, Tissue recombination | ↓Bmi-1 and ↑FGF10→↓graft size, ↓tumorigenic phenotype relative to FGF10 overexpression alone | Lukacs et al. (2010) |

including proliferation, apoptosis, differentiation, and inflammatory responses. *MAP3K7* is hemizygously deleted in 30–40% of primary prostate tumors and its loss is correlated with higher-grade cancers (Liu et al. 2007). To validate TAK1 as a prostate tumor suppressor, we suppressed Tak1 expression in mouse PrP/SCs and characterized these cells in vitro and in vivo. Tak1 suppression altered cell morphology and significantly increased proliferation, migration, and invasion of PrP/SCs. In tissue recombination experiments using control or Tak1 shRNA PrP/SCs mixed with rat UGM, the phenotype of Tak1-suppressed grafts was much different than control grafts. After 10 weeks of growth, grafts formed from control cells underwent complete lineage differentiation and generated benign prostatic structures, while Tak1-suppressed grafts formed heterogenous structures with benign, PIN, and invasive carcinoma phenotypes represented. Furthermore, loss of Tak1 promoted cellular proliferation in vivo as determined by increased expression of Ki67 in Tak1-suppressed grafts relative to control grafts (Wu et al. 2012). Utilization of the PrP/SC and tissue recombination models allowed us to identify a novel and critical prostate tumor suppressor, TAK1, which may serve as a diagnostic biomarker or therapeutic target for treatment of prostate cancer.

**PIM1**, a serine/threonine kinase involved in cell cycle regulation and inhibition of apoptosis, is overexpressed in high-grade PIN and prostate cancer (Cibull et al. 2006; Dhanasekaran et al. 2001). In unpublished studies in collaboration with Michael Lilly, we evaluated PIM1 as an oncogene in tumor development. We created stable clonal mouse PrP/SC lines that overexpressed Pim1 and assessed the phenotypic consequences. Overexpression of Pim1 did not affect monolayer growth, but did significantly increase 3D growth in collagen. Control or Pim1-overexpressing PrP/SCs were then recombined with rat UGM, grafted under the renal capsules of immunocompromised mice, and removed after 12 weeks. While grafts from control PrP/SC tissue recombinants formed normal ductal structures, Pim1-PrP/SC recombinants displayed a phenotype consistent with high-grade PIN. Pim1-PrP/SC grafts retained intact p63+ basal and AR+ luminal layers but formed abnormal structures characterized by crowded, irregularly spaced micropapillary epithelial cells (Chen, Lilly and Cramer, unpublished). Therefore, Pim1 kinase plays a role in formation of PIN but is insufficient to induce prostate tumorigenesis. Since Pim1-PrP/SCs are in an "initiated" state, further genetic insult may lead to tumor development in this model. In a recent study by Abdulkadir and colleagues, Pim1 was found to cooperate with c-MYC to form highly vascularized carcinomas in tissue recombinants grafted under the renal capsule for 6 weeks (Wang et al. 2010). Collectively, these data demonstrate that Pim1 has an oncogenic role in prostate cancer development.

## 9.7 Discussion and Future Considerations

The complex mutational landscape of the prostate cancer genome makes identification of critical regulators of prostate tumorigenesis very challenging. Because numerous genomic alterations exist within a primary prostate tumor, the major goals of prostate cancer research today are to (1) distinguish driver oncogenes/tumor suppressors genes from passenger genes, (2) understand how multiple oncogenes and tumor suppressors cooperate to initiate and progress tumor development, and (3) identify new prognostic indicators and therapeutic targets to improve survival for patients with aggressive disease. The stem cell models described here have been used effectively to identify critical prostate oncogenes and tumor suppressors including PTEN, RAS, P53, and PIM1 as well as others listed in Table 9.1. Some models have begun to explore cooperativity among multiple cancer genes and to test therapeutics, but these types of studies are still in the early stages. Although development of a "perfect" stem cell model may never be achievable, current models can be improved upon to more accurately mimic the genetic events that contribute to tumor initiation, progression, and metastasis. Considerations for future models should include:

1. *Promoter specificity*
   Identification of a prostate stem cell-specific promoter would greatly enhance current models. In other tissues, transgenic mice have been developed using tissue-specific stem cell genes to drive target expression. For example, Prx1 and Dermo1 promoters have been used to regulate gene expression in mesenchymal stem cells which are precursors for bone development (Elefteriou and Yang 2011). To date, the only epithelial prostate-specific promoters utilized in GEM models have been from probasin, PSA, and Nkx3.1, genes known to be highly expressed in more differentiated cells of the prostate (i.e., luminal cells). The FSP1 promoter has been used to drive Cre expression in prostatic fibroblasts (Bhowmick et al. 2004). However, recently, Tang and colleagues used the PSA promoter in a unique way to identify a castration-resistant stem/progenitor population. In this study, LNCaP cells were infected with a lentiviral PSAP-GFP construct in which the PSA promoter drives expression of eGFP. FACS was then used to isolate the brightest 10% of GFP-expressing cells (GFP+) and the lowest 2–6% of cells expressing GFP ($GFP^{-/lo}$). GFP level correlated with PSA expression in these cell fractions (i.e., GFP+ = PSA+, $GFP^{-/lo}$ = $PSA^{-/lo}$). Gene expressing profiling revealed that $PSA^{-/lo}$ LNCaP cells overexpress numerous genes involved in antistress signaling and DNA damage repair but have lower expression of antiapoptotic, cell cycle, and mitosis genes. This profile is suggestive of quiescence and stress resistance—characteristics displayed by stem cells. Indeed, $PSA^{-/lo}$ LNCaP cells possess high colony and prostasphere-forming abilities, can undergo asymmetric cell division, are resistant to androgen deprivation, and express stem cell-associated genes. Furthermore, $PSA^{-/lo}$ LAPC9 xenografts maintain tumorigenicity upon serial transplantation and are resistant to androgen deprivation therapy in vivo (Qin et al. 2012). Future studies could utilize this model to alter gene expression in $PSA^{+}$ and $PSA^{-/lo}$ subpopulations in different cell lines and assess tumorigenicity in vivo. Cell-specific models can also be useful for modeling the roles of genomic changes in prostate tumorgenesis. The CK14 and CK8 promoters have been used to drive loss of Pten in basal and luminal cells, respectively (Choi et al. 2012). However, a limitation of this model is that Pten is deleted in all cells expressing CK14 or CK8 not just prostate cells. This model could

be adapted to study gene alterations in a prostate and cell-specific manner by isolation of CK14-Pten and CK8-Pten PrP/SCs and generation of tissue recombinants from these different cell types.

2. *Regulated compound models*
From the numerous models described in this chapter, it is clear that one gene alone cannot drive tumor initiation, progression, and metastasis. Likely, prostate tumorigenesis occurs when multiple alterations occur in critical signaling pathways and override DNA damage and cell cycle checkpoint responses. Future compound models should examine three or more alterations in combination since current two-hit models do not adequately recapitulate metastatic prostate cancer. Since the RB, AKT/PI3K, and RAS/MAPK pathways are most frequently misregulated in prostate tumorigenesis (Taylor et al. 2010), a triple model of Rb loss, Pten loss, and Ras overexpression could result in a more aggressive metastatic phenotype than that observed in the compound Pten$^{-/-}$/activated Ras model (Mulholland et al. 2012). Combined p53 loss, Ras overexpression, and Myc overexpression may also be a potentially aggressive model since early experiments using UGS demonstrated these three hits together resulted in bone metastasis (Thompson et al. 1995). Another strategy that should be utilized when developing future models is the use of regulated genetics. A weakness of most current models is that genetic alterations take effect before, during, or after prostatic development in GEM and tissue recombinants at specific time points that cannot be altered. If gene activation or deletion could be temporally regulated, then these alterations could be introduced into fully mature adult mice with developed prostates or fully formed prostatic glandular structures of tissue recombinants. Furthermore, a critical question in prostate cancer research continues to be do genetic alterations arise in a particular order during initiation and progression of tumorigenesis? Different systems of regulation could be used in combination to evaluate multiple oncogenes and tumor suppressors in a controlled manner. Current systems that use regulated genetics to control gene expression include the ER-Cre, Tet-ON/OFF, Ecdysone, and FLP-FRT systems (Feil et al. 1996; Galimi et al. 2005; Gossen and Bujard 1992; Sadowski 1995).

3. *Microenvironment*
The tumor microenvironment can significantly impact tumor initiation, progression, and metastasis. Different types of inflammatory and immune cells, for example, can positively or negatively affect tumor growth and invasion through cross talk with tumor cells. Historically, rat UGM is used in tissue recombination models so grafts must be implanted into immunocompromised hosts. In future studies, the use of syngeneic models (e.g., C57BL/6 UGM+C57BL/6 PrP/SCs) in animals with intact immune systems (i.e., C57BL/6) will better recapitulate tumor development in human patients. Additionally, epithelial-stromal interactions play a critical role in both normal prostatic development and tumorigenesis (Cunha et al. 2003). For example, overexpression of FGF10 in the mesenchyme can initiate prostate carcinoma (Lawson et al. 2010; Memarzadeh et al. 2007). As other stromal regulators of tumorigenesis are identified, they can be evaluated using a similar

approach. Eventually, compound models can be developed in which different combinations of genetic alterations are introduced into the stromal and epithelial compartments of the prostate.

4. *Metastasis*
   Advanced prostate cancer primarily metastasizes to the bone, yet current models have failed to reliably generate bone lesions. In a recent compound model of Pten loss and Kras activation, cells with the PB-$Cre^{+}Pten^{L/W}Kras^{G12D/W}$ and PB-$Cre^{+}Pten^{L/L}Kras^{G12D/W}$ genotypes were identified in bone marrow flushes from GEM mice, but due to early lethality, bone metastases were not observed (Mulholland et al. 2012). As we identify more genes involved in prostate cancer progression, castration resistance, and metastasis, stem cell models can be used to assess the ability of these genes to impact development of bone metastases. The utilization of compound genomic models with $\geq$3 alterations will likely produce more aggressive tumors with a greater potential for metastasis.

5. *Human primary cells*
   Most current stem cell models utilize mouse cells to evaluate candidate prostate cancer genes. However, two major differences exist between the mouse and human prostates: (1) the anatomies are different and (2) mice do not spontaneously develop prostate cancer without genetic manipulation. Therefore, stem cell models that use normal and tumorigenic human primary prostate cells may be better tools for validating putative oncogenes and tumor suppressors. Currently, epithelial cells can be isolated from primary tumors, stem cell populations can be enriched using FACS, genetics can be manipulated, and cells can be combined with rat UGM in tissue recombinants grafted into immunocompromised mice (Goldstein et al. 2011). This approach has been used successfully to demonstrate cooperativity of oncogenic AKT, ERG, and AR (Goldstein et al. 2010). However, this method has limitations because primary samples are not cultured and expanded so experiments cannot be repeated to verify results. As strategies for human PrP/SC culture improve, primary cell models will become more important for validation of candidate cancer genes.

**Acknowledgements** The authors thank the following individuals for their scientific contributions to this chapter: Lina Romero for FACS analysis and characterization of PrP/SCs from different genetic backgrounds; Molishree Joshi for isolation of $Tak1^{L/L}$ MPECs and microscopy; Wenhong Chen and Michael Lilly for Pim1 studies.

## References

Abou-Kheir WG, Hynes PG, Martin PL, Pierce R, Kelly K (2010) Characterizing the contribution of stem/progenitor cells to tumorigenesis in the $Pten^{-/-}$ $TP53^{-/-}$ prostate cancer model. Stem Cells 28:2129–2140

Abou-Kheir W, Hynes PG, Martin P, Yin JJ, Liu YN, Seng V, Lake R, Spurrier J, Kelly K (2011) Self-renewing $Pten^{-/-}$ $TP53^{-/-}$ protospheres produce metastatic adenocarcinoma cell lines with multipotent progenitor activity. PLoS One 6:e26112

Axanova LS, Chen YQ, McCoy T, Sui G, Cramer SD (2010) 1,25-dihydroxyvitamin D(3) and PI3K/AKT inhibitors synergistically inhibit growth and induce senescence in prostate cancer cells. Prostate 70:1658–1671

Barclay WW, Cramer SD (2005) Culture of mouse prostatic epithelial cells from genetically engineered mice. Prostate 63:291–298

Barclay WW, Axanova LS, Chen W, Romero L, Maund SL, Soker S, Lees CJ, Cramer SD (2008) Characterization of adult prostatic progenitor/stem cells exhibiting self-renewal and multilineage differentiation. Stem Cells 26:600–610

Berger MF, Lawrence MS, Demichelis F, Drier Y, Cibulskis K, Sivachenko AY, Sboner A, Esgueva R, Pflueger D, Sougnez C, Onofrio R, Carter SL, Park K, Habegger L, Ambrogio L, Fennell T, Parkin M, Saksena G, Voet D, Ramos AH, Pugh TJ, Wilkinson J, Fisher S, Winckler W, Mahan S, Ardlie K, Baldwin J, Simons JW, Kitabayashi N, MacDonald TY, Kantoff PW, Chin L, Gabriel SB, Gerstein MB, Golub TR, Meyerson M, Tewari A, Lander ES, Getz G, Rubin MA, Garraway LA (2011) The genomic complexity of primary human prostate cancer. Nature 470:214–220

Bhowmick NA, Neilson EG, Moses HL (2004) Stromal fibroblasts in cancer initiation and progression. Nature 432:332–337

Carver BS, Tran J, Gopalan A, Chen Z, Shaikh S, Carracedo A, Alimonti A, Nardella C, Varmeh S, Scardino PT, Cordon-Cardo C, Gerald W, Pandolfi PP (2009) Aberrant ERG expression cooperates with loss of PTEN to promote cancer progression in the prostate. Nat Genet 41:619–624

Casey OM, Fang L, Hynes PG, Abou-Kheir WG, Martin PL, Tillman HS, Petrovics G, Awwad HO, Ward Y, Lake R, Zhang L, Kelly K (2012) TMPRSS2- driven ERG expression in vivo increases self-renewal and maintains expression in a castration resistant subpopulation. PLoS One 7:e41668

Chiaverotti T, Couto SS, Donjacour A, Mao JH, Nagase H, Cardiff RD, Cunha GR, Balmain A (2008) Dissociation of epithelial and neuroendocrine carcinoma lineages in the transgenic adenocarcinoma of mouse prostate model of prostate cancer. Am J Pathol 172:236–246

Choi N, Zhang B, Zhang L, Ittmann M, Xin L (2012) Adult murine prostate basal and luminal cells are self-sustained lineages that can both serve as targets for prostate cancer initiation. Cancer Cell 21:253–265

Cibull TL, Jones TD, Li L, Eble JN, Ann Baldridge L, Malott SR, Luo Y, Cheng L (2006) Overexpression of Pim-1 during progression of prostatic adenocarcinoma. J Clin Pathol 59:285–288

Clarke AR, Maandag ER, van Roon M, van der Lugt NM, van der Valk M, Hooper ML, Berns A, te Riele H (1992) Requirement for a functional Rb-1 gene in murine development. Nature 359:328–330

Collins AT, Habib FK, Maitland NJ, Neal DE (2001) Identification and isolation of human prostate epithelial stem cells based on alpha(2)beta(1)-integrin expression. J Cell Sci 114:3865–3872

Collins AT, Berry PA, Hyde C, Stower MJ, Maitland NJ (2005) Prospective identification of tumorigenic prostate cancer stem cells. Cancer Res 65:10946–10951

Cunha GR (1972a) Epithelio-mesenchymal interactions in primordial gland structures which become responsive to androgenic stimulation. Anat Rec 172:179–195

Cunha GR (1972b) Tissue interactions between epithelium and mesenchyme of urogenital and integumental origin. Anat Rec 172:529–541

Cunha GR, Hayward SW, Wang YZ, Ricke WA (2003) Role of the stromal microenvironment in carcinogenesis of the prostate. Int J Cancer 107:1–10

Dhanasekaran SM, Barrette TR, Ghosh D, Shah R, Varambally S, Kurachi K, Pienta KJ, Rubin MA, Chinnaiyan AM (2001) Delineation of prognostic biomarkers in prostate cancer. Nature 412:822–826

Dubrovska A, Kim S, Salamone RJ, Walker JR, Maira SM, Garcia-Echeverria C, Schultz PG, Reddy VA (2009) The role of PTEN/Akt/PI3K signaling in the maintenance and viability of prostate cancer stem-like cell populations. Proc Natl Acad Sci USA 106:268–273

Elefteriou F, Yang X (2011) Genetic mouse models for bone studies—strengths and limitations. Bone 49:1242–1254

English HF, Santen RJ, Isaacs JT (1987) Response of glandular versus basal rat ventral prostatic epithelial cells to androgen withdrawal and replacement. Prostate 11:229–242
Evans GS, Chandler JA (1987) Cell proliferation studies in the rat prostate: II. The effects of castration and androgen-induced regeneration upon basal and secretory cell proliferation. Prostate 11:339–351
Feil R, Brocard J, Mascrez B, LeMeur M, Metzger D, Chambon P (1996) Ligand-activated site-specific recombination in mice. Proc Natl Acad Sci USA 93:10887–10890
Galimi F, Saez E, Gall J, Hoong N, Cho G, Evans RM, Verma IM (2005) Development of ecdysone-regulated lentiviral vectors. Mol Ther 11:142–148
Goldstein AS, Huang J, Guo C, Garraway IP, Witte ON (2010) Identification of a cell of origin for human prostate cancer. Science 329:568–571
Goldstein AS, Drake JM, Burnes DL, Finley DS, Zhang H, Reiter RE, Huang J, Witte ON (2011) Purification and direct transformation of epithelial progenitor cells from primary human prostate. Nat Protoc 6:656–667
Goodyear SM, Amatangelo MD, Stearns ME (2009) Dysplasia of human prostate CD133(hi) sub-population in NOD-SCIDS is blocked by c-myc anti-sense. Prostate 69:689–698
Gossen M, Bujard H (1992) Tight control of gene expression in mammalian cells by tetracycline-responsive promoters. Proc Natl Acad Sci USA 89:5547–5551
Grasso CS, Wu YM, Robinson DR, Cao X, Dhanasekaran SM, Khan AP, Quist MJ, Jing X, Lonigro RJ, Brenner JC, Asangani IA, Ateeq B, Chun SY, Siddiqui J, Sam L, Anstett M, Mehra R, Prensner JR, Palanisamy N, Ryslik GA, Vandin F, Raphael BJ, Kunju LP, Rhodes DR, Pienta KJ, Chinnaiyan AM, Tomlins SA (2012) The mutational landscape of lethal castration-resistant prostate cancer. Nature 487:239–243
Greaves M, Maley CC (2012) Clonal evolution in cancer. Nature 481:306–313
Imagawa W, Tomooka Y, Nandi S (1982) Serum-free growth of normal and tumor mouse mammary epithelial cells in primary culture. Proc Natl Acad Sci USA 79:4074–4077
Isaacs JT, Coffey DS (1989) Etiology and disease process of benign prostatic hyperplasia. Prostate Suppl 2:33–50
Ishkanian AS, Mallof CA, Ho J, Meng A, Albert M, Syed A, van der Kwast T, Milosevic M, Yoshimoto M, Squire JA, Lam WL, Bristow RG (2009) High-resolution array CGH identifies novel regions of genomic alteration in intermediate-risk prostate cancer. Prostate 69:1091–1100
Jacks T, Fazeli A, Schmitt EM, Bronson RT, Goodell MA, Weinberg RA (1992) Effects of an Rb mutation in the mouse. Nature 359:295–300
Jin RJ, Lho Y, Wang Y, Ao M, Revelo MP, Hayward SW, Wills ML, Logan SK, Zhang P, Matusik RJ (2008) Down-regulation of p57Kip2 induces prostate cancer in the mouse. Cancer Res 68:3601–3608
King JC, Xu J, Wongvipat J, Hieronymus H, Carver BS, Leung DH, Taylor BS, Sander C, Cardiff RD, Couto SS, Gerald WL, Sawyers CL (2009) Cooperativity of TMPRSS2-ERG with PI3-kinase pathway activation in prostate oncogenesis. Nat Genet 41:524–526
Klezovitch O, Risk M, Coleman I, Lucas JM, Null M, True LD, Nelson PS, Vasioukhin V (2008) A causal role for ERG in neoplastic transformation of prostate epithelium. Proc Natl Acad Sci USA 105:2105–2110
Kubota K, Preislef HD, Lok MS, Minowada J (1981) Lack of effect of colony-stimulating activity on human myeloid leukemia cell line (ML-2) cells. Leuk Res 5:311–320
Kusama Y, Enami J, Kano Y (1989) Growth and morphogenesis of mouse prostate epithelial cells in collagen gel matrix culture. Cell Biol Int Rep 13:569–575
Lawson DA, Zong Y, Memarzadeh S, Xin L, Huang J, Witte ON (2010) Basal epithelial stem cells are efficient targets for prostate cancer initiation. Proc Natl Acad Sci USA 107:2610–2615
Lee EY, Chang CY, Hu N, Wang YC, Lai CC, Herrup K, Lee WH, Bradley A (1992) Mice deficient for Rb are nonviable and show defects in neurogenesis and haematopoiesis. Nature 359:288–294
Lei Q, Jiao J, Xin L, Chang CJ, Wang S, Gao J, Gleave ME, Witte ON, Liu X, Wu H (2006) NKX3.1 stabilizes p53, inhibits AKT activation, and blocks prostate cancer initiation caused by PTEN loss. Cancer Cell 9:367–378

Li H, Chen X, Calhoun-Davis T, Claypool K, Tang DG (2008) PC3 human prostate carcinoma cell holoclones contain self-renewing tumor-initiating cells. Cancer Res 68:1820–1825

Liu W, Chang BL, Cramer S, Koty PP, Li T, Sun J, Turner AR, Von Kap-Herr C, Bobby P, Rao J, Zheng SL, Isaacs WB, Xu J (2007) Deletion of a small consensus region at 6q15, including the MAP3K7 gene, is significantly associated with high-grade prostate cancers. Clin Cancer Res 13:5028–5033

Liu W, Xie CC, Zhu Y, Li T, Sun J, Cheng Y, Ewing CM, Dalrymple S, Turner AR, Isaacs JT, Chang BL, Zheng SL, Isaacs WB, Xu J (2008) Homozygous deletions and recurrent amplifications implicate new genes involved in prostate cancer. Neoplasia 10:897–907

Liu J, Pascal LE, Isharwal S, Metzger D, Ramos Garcia R, Pilch J, Kasper S, Williams K, Basse PH, Nelson JB, Chambon P, Wang Z (2011) Regenerated luminal epithelial cells are derived from preexisting luminal epithelial cells in adult mouse prostate. Mol Endocrinol 25:1849–1857

Liu W, Lindberg J, Sui G, Luo J, Egevad L, Li T, Xie C, Wan M, Kim ST, Wang Z, Turner AR, Zhang Z, Feng J, Yan Y, Sun J, Bova GS, Ewing CM, Yan G, Gielzak M, Cramer SD, Vessella RL, Zheng SL, Gronberg H, Isaacs WB, Xu J (2012) Identification of novel CHD1-associated collaborative alterations of genomic structure and functional assessment of CHD1 in prostate cancer. Oncogene 31:3939–3948

Luchman HA, Friedman HC, Villemaire ML, Peterson AC, Jirik FR (2008) Temporally controlled prostate epithelium-specific gene alterations. Genesis 46:229–234

Lukacs RU, Memarzadeh S, Wu H, Witte ON (2010) Bmi-1 is a crucial regulator of prostate stem cell self-renewal and malignant transformation. Cell Stem Cell 7:682–693

Maddison LA, Sutherland BW, Barrios RJ, Greenberg NM (2004) Conditional deletion of Rb causes early stage prostate cancer. Cancer Res 64:6018–6025

Maund SL, Barclay WW, Hover LD, Axanova LS, Sui G, Hipp JD, Fleet JC, Thorburn A, Cramer SD (2011) Interleukin-1alpha mediates the antiproliferative effects of 1,25-dihydroxyvitamin D3 in prostate progenitor/stem cells. Cancer Res 71:5276–5286

McKeehan WL, Adams PS, Rosser MP (1982) Modified nutrient medium MCDB 151, defined growth factors, cholera toxin, pituitary factors, and horse serum support epithelial cell and suppress fibroblast proliferation in primary cultures of rat ventral prostate cells. In Vitro 18:87–91

Memarzadeh S, Xin L, Mulholland DJ, Mansukhani A, Wu H, Teitell MA, Witte ON (2007) Enhanced paracrine FGF10 expression promotes formation of multifocal prostate adenocarcinoma and an increase in epithelial androgen receptor. Cancer Cell 12:572–585

Mulholland DJ, Xin L, Morim A, Lawson D, Witte O, Wu H (2009) Lin-Sca-1+CD49fhigh stem/progenitors are tumor-initiating cells in the Pten-null prostate cancer model. Cancer Res 69:8555–8562

Mulholland DJ, Kobayashi N, Ruscetti M, Zhi A, Tran LM, Huang J, Gleave M, Wu H (2012) Pten loss and RAS/MAPK activation cooperate to promote EMT and metastasis initiated from prostate cancer stem/progenitor cells. Cancer Res 72:1878–1889

Patrawala L, Calhoun T, Schneider-Broussard R, Li H, Bhatia B, Tang S, Reilly JG, Chandra D, Zhou J, Claypool K, Coghlan L, Tang DG (2006) Highly purified CD44+ prostate cancer cells from xenograft human tumors are enriched in tumorigenic and metastatic progenitor cells. Oncogene 25:1696–1708

Peehl DM, Stamey TA (1984) Serial propagation of adult human prostatic epithelial cells with cholera toxin. In Vitro 20:981–986

Peehl DM, Stamey TA (1986) Growth responses of normal, benign hyperplastic, and malignant human prostatic epithelial cells in vitro to cholera toxin, pituitary extract, and hydrocortisone. Prostate 8:51–61

Qin J, Liu X, Laffin B, Chen X, Choy G, Jeter CR, Calhoun-Davis T, Li H, Palapattu GS, Pang S, Lin K, Huang J, Ivanov I, Li W, Suraneni MV, Tang DG (2012) The PSA(−/lo) prostate cancer cell population harbors self-renewing long-term tumor-propagating cells that resist castration. Cell Stem Cell 10:556–569

Ratnacaram CK, Teletin M, Jiang M, Meng X, Chambon P, Metzger D (2008) Temporally controlled ablation of PTEN in adult mouse prostate epithelium generates a model of invasive prostatic adenocarcinoma. Proc Natl Acad Sci USA 105:2521–2526

Robbins CM, Tembe WA, Baker A, Sinari S, Moses TY, Beckstrom-Sternberg S, Beckstrom-Sternberg J, Barrett M, Long J, Chinnaiyan A, Lowey J, Suh E, Pearson JV, Craig DW, Agus DB, Pienta KJ, Carpten JD (2011) Copy number and targeted mutational analysis reveals novel somatic events in metastatic prostate tumors. Genome Res 21:47–55

Sadowski PD (1995) The Flp recombinase of the 2-microns plasmid of Saccharomyces cerevisiae. Prog Nucleic Acid Res Mol Biol 51:53–91

Scherl A, Li JF, Cardiff RD, Schreiber-Agus N (2004) Prostatic intraepithelial neoplasia and intestinal metaplasia in prostates of probasin-RAS transgenic mice. Prostate 59:448–459

Shaw A, Papadopoulos J, Johnson C, Bushman W (2006) Isolation and characterization of an immortalized mouse urogenital sinus mesenchyme cell line. Prostate 66:1347–1358

Shen MM, Abate-Shen C (2010) Molecular genetics of prostate cancer: new prospects for old challenges. Genes Dev 24:1967–2000

Shi X, Gipp J, Bushman W (2007) Anchorage-independent culture maintains prostate stem cells. Dev Biol 312:396–406

Solimini NL, Xu Q, Mermel CH, Liang AC, Schlabach MR, Luo J, Burrows AE, Anselmo AN, Bredemeyer AL, Li MZ, Beroukhim R, Meyerson M, Elledge SJ (2012) Recurrent hemizygous deletions in cancers may optimize proliferative potential. Science 337:104–109

Taylor BS, Schultz N, Hieronymus H, Gopalan A, Xiao Y, Carver BS, Arora VK, Kaushik P, Cerami E, Reva B, Antipin Y, Mitsiades N, Landers T, Dolgalev I, Major JE, Wilson M, Socci ND, Lash AE, Heguy A, Eastham JA, Scher HI, Reuter VE, Scardino PT, Sander C, Sawyers CL, Gerald WL (2010) Integrative genomic profiling of human prostate cancer. Cancer Cell 18:11–22

Thompson TC, Southgate J, Kitchener G, Land H (1989) Multistage carcinogenesis induced by ras and myc oncogenes in a reconstituted organ. Cell 56:917–930

Thompson TC, Park SH, Timme TL, Ren C, Eastham JA, Donehower LA, Bradley A, Kadmon D, Yang G (1995) Loss of p53 function leads to metastasis in ras+myc-initiated mouse prostate cancer. Oncogene 10:869–879

Timme T L, Yang G, Rogers E, Kadmon D, Morganstern J P, Park S H, Thompson T C (1996) Retroviral transduction of transforming growth factor-beta1 induces pleiotropic benign prostatic growth abnormalities in mouse prostate reconstitutions. Lab Invest 74: 747–760

Tomlins SA, Laxman B, Varambally S, Cao X, Yu J, Helgeson BE, Cao Q, Prensner JR, Rubin MA, Shah RB, Mehra R, Chinnaiyan AM (2008) Role of the TMPRSS2-ERG gene fusion in prostate cancer. Neoplasia 10:177–188

Tran CP, Lin C, Yamashiro J, Reiter RE (2002) Prostate stem cell antigen is a marker of late intermediate prostate epithelial cells. Mol Cancer Res 1:113–121

van Bokhoven A, Varella-Garcia M, Korch C, Johannes WU, Smith EE, Miller HL, Nordeen SK, Miller GJ, Lucia MS (2003) Molecular characterization of human prostate carcinoma cell lines. Prostate 57:205–225

Wang Y, Hayward SW, Donjacour AA, Young P, Jacks T, Sage J, Dahiya R, Cardiff RD, Day ML, Cunha GR (2000) Sex hormone-induced carcinogenesis in Rb-deficient prostate tissue. Cancer Res 60:6008–6017

Wang S, Garcia AJ, Wu M, Lawson DA, Witte ON, Wu H (2006) Pten deletion leads to the expansion of a prostatic stem/progenitor cell subpopulation and tumor initiation. Proc Natl Acad Sci USA 103:1480–1485

Wang X, Kruithof-de Julio M, Economides KD, Walker D, Yu H, Halili MV, Hu YP, Price SM, Abate-Shen C, Shen MM (2009) A luminal epithelial stem cell that is a cell of origin for prostate cancer. Nature 461:495–500

Wang J, Kim J, Roh M, Franco OE, Hayward SW, Wills ML, Abdulkadir SA (2010) Pim1 kinase synergizes with c-MYC to induce advanced prostate carcinoma. Oncogene 29:2477–2487

Wicha MS, Liu S, Dontu G (2006) Cancer stem cells: an old idea—a paradigm shift. Cancer Res 66:1883–1890. discussion 1895–1886

Wu M, Shi L, Cimic A, Romero L, Sui G, Lees CJ, Cline JM, Seals DF, Sirintrapun JS, McCoy TP, Liu W, Kim JW, Hawkins GA, Peehl DM, Xu J, Cramer SD (2012) Suppression of Tak1 promotes prostate tumorigenesis. Cancer Res 72:2833–2843

Xin L, Lawson DA, Witte ON (2005) The Sca-1 cell surface marker enriches for a prostate-regenerating cell subpopulation that can initiate prostate tumorigenesis. Proc Natl Acad Sci USA 102:6942–6947

Xin L, Lukacs RU, Lawson DA, Cheng D, Witte ON (2007) Self-renewal and multilineage differentiation in vitro from murine prostate stem cells. Stem Cells 25:2760–2769

Yang J, Guzman R, Richards J, Imagawa W, McCormick K, Nandi S (1980) Growth factor- and cyclic nucleotide-induced proliferation of normal and malignant mammary epithelial cells in primary culture. Endocrinology 107:35–41

Zong Y, Xin L, Goldstein AS, Lawson DA, Teitell MA, Witte ON (2009) ETS family transcription factors collaborate with alternative signaling pathways to induce carcinoma from adult murine prostate cells. Proc Natl Acad Sci USA 106:12465–12470

# Index



S.D. Cramer (ed.), *Stem Cells and Prostate Cancer*, DOI 10.1007/978-1-4614-6498-3,

MIX
Papier aus verantwortungsvollen Quellen
Paper from responsible sources
FSC® C105338

If you have any concerns about our products,
you can contact us on
**ProductSafety@springernature.com**

In case Publisher is established outside the EU,
the EU authorized representative is:
**Springer Nature Customer Service Center GmbH**
**Europaplatz 3, 69115 Heidelberg, Germany**

Printed by Libri Plureos GmbH
in Hamburg, Germany